BRITISH GEOLOGICAL SURVEY
Natural Environment Research Council

R. W. GALLOIS

Geology of the country around Ely

Memoir for 1:50 000 geological sheet 173 (England and Wales)

CONTRIBUTORS

Palaeontology
Beris M. Cox
H. C. Ivimey-Cook
A. A. Morter

Soils and Land Use
R. S. Seale

LONDON: HER MAJESTY'S STATIONERY OFFICE   1988

First published 1988

ISBN 0 11 884395 8

Bibliographical reference
GALLOIS, R. W. 1988. Geology of the country around Ely. *Mem. Br. Geol. Surv.*, Sheet 173.

Author
R. W. GALLOIS, BSc, DIC, PhD, CEng, FIMM
British Geological Survey, Edinburgh

Contributors
B. M. Cox, BSc, PhD
H. C. Ivimey-Cook, BSc, PhD
British Geological Survey, Keyworth

A. A. Morter, BSc
formerly of *British Geological Survey, Keyworth*

R. S. Seale, BSc
Soil Survey of England and Wales,
Brooklands Avenue, Cambridge

Other publications of the Survey dealing with this district and adjoining districts

BOOKS

Memoirs
Geology of the country around Cambridge, Sheet 188, 1969
Geology of the country around Bury St Edmunds, Sheet 189, in preparation
Geology of the country around King's Lynn, Sheets 129 and 145, in preparation
Geology of the country around Huntingdon and Biggleswade, Sheets 187 and 204, 1965

British Regional Geology
East Anglia and adjoining areas (4th edition), 1961

MAPS

1:50 000 (Solid and Drift)
Sheet 129/145   King's Lynn and the Wash, 1978
Sheet 158   Peterborough, 1985
Sheet 187   Huntingdon, 1975
Sheet 188   Cambridge, 1981
Sheet 189   Bury St Edmunds, 1982

1:250 000
Sheet 52N 00   East Anglia, solid geology, 1986
Sheet 52N 00   East Anglia, aeromagnetic anomaly
Sheet 52N 00   East Anglia, Bouguer gravity anomaly

1:253 440 (four miles to one inch) (Solid and Drift)
Sheet 12   Louth, Peterborough, Norwich and
                 Yarmouth, 1986

1:125 000 Hydrogeology
Sheet 4   Northern East Anglia (two sheets), 1976

Printed in the United Kingdom for Her Majesty's Stationery Office

Dd 0240401 5/88 C20 398 12521

Geology of the country around Ely

Few areas of Britain have undergone as great a transformation as that experienced by the Ely district during the past 200 years. A landscape of fens, shallow meres and winding creeks, with islands of only slightly higher ground, has been converted into an almost treeless arable plain cut by a complex network of long straight drains. Although these dramatic changes have been man-made, they have merely accentuated geological features that have controlled this scenery for more than a million years: features which themselves reflect a succession of environments that were present in the distant geological past. This memoir traces the geological history of the district from the muddy seas of the Silurian of 400 million years ago, via the deserts of the Triassic and the shallow tropical and subtropical seas of the Jurassic and Cretaceous, to the Quaternary glaciations of 250 000 to 18 000 years ago when the district was covered by ice sheets up to several thousand feet thick. Finally, it records the history of the period after the retreat of the ice when temperate climates returned and the broad glaciated hollow that was to become Fenland was infilled with muds and peats.

The oldest rocks at surface in the Ely district are Jurassic mudstones with a rich and varied fauna, including ammonites, bivalves, gastropods and marine reptiles, that shows them to have been deposited in a warm shallow sea, probably close to land. These readily erodible rocks underlie the vast embayment of low ground that now forms Fenland and The Wash. Their deposition was succeeded by that of shallow-water marine sands, the Jurassic Sandringham Sands and the Cretaceous Lower Greensand, which although themselves only poorly indurated are sufficiently strong to cap the islands on which Ely, Southery and Stuntney are built. In the eastern part of the district the soft marine Cretaceous limestones of the Chalk form an escarpment that marks the limit of Fenland. Glacial deposits, mostly stony clays derived from the Jurassic rocks and the Chalk, and meltwater gravels composed largely of flints, cover relatively small areas of the district. In the absence of hard rocks they have an important influence on the scenery and cap the islands and higher ground at Downham, Littleport, Manea, Pymore, Stonea, Sutton, Wimblington and Witchford.

The greater part of the district is occupied by the Recent sediments of Fenland. These were deposited during the past 10 000 years, during which period sea level has risen by about 30 m. Through man's activities the fens have been converted to richly fertile peat soils and the meres to calcareous marls. But in the process, shrinkage and wastage of the soils due to the drainage works have caused large areas of the underlying geology, including the Jurassic clays, Quaternary gravels and Recent marine clays, to be revealed and have given rise to markedly less fertile soils. The future planning of the economic well-being of the district must, therefore, be based on a detailed knowledge of the geology, its relationship to the soils and to the past and present drainage patterns.

This memoir provides an authoritative account of all aspects of the geology of the Ely district including descriptions of the rock types and their faunas, their structure and geological history, and the relationships of these features to past and present land use.

Frontispiece River Great Ouse looking northwards from Brandon Creek towards Southery. The river flows between high banks that protect the surrounding low ground of Southery Fens from flooding. This part of the river was probably originally the headwaters of a large tidal creek that entered the sea at King's Lynn. It became considerably enlarged in Roman times when the main waters of the Great Ouse were diverted from Littleport (and their outfall at Wisbech) to Brandon Creek (Aerofilms)

CONTENTS

TABLES

PLATES

PREFACE

This memoir is an explanatory account to accompany the provisional edition of 1 to 50 000 scale geological Sheet 173 (Ely), published in 1980. The map has been prepared as a collaborative effort between Mr R. S. Seale of the Soil Survey of England and Wales and Dr R. W. Gallois of the East Anglia and South-Eastern England Unit of the Geological Survey. It is based on Mr Seale's six inch-to-the-mile maps of the soils of the area, published at the one-inch scale as Soil Survey Sheet 173 in 1972, together with field data collected from temporary sections and boreholes and collated by Dr Gallois. The present account describes the geology of the district and brings together the information used in preparing the geological map.

The original geological survey of the southern part of the district was at the one-inch scale by F. J. Bennett, A. J. Jukes-Browne, H. W. Penning, S. B. J. Skertchly and H. B. Woodward (Sheet 51, published 1881–83) and the northern part by C. Bristow, F. J. Bennett, A. C. Cameron, C. E. Hawkins, C. Reid, S. B. J. Skertchly, W. Whitaker and H. B. Woodward (Sheet 65, published 1886). The accompanying memoirs to these sheets, *The geology of Ely, Mildenhall and Thetford* (Whitaker, Woodward and Jukes-Browne, 1891) and *The geology of Norfolk, south-western and northern Cambridgeshire* (Whitaker, Skertchly and Jukes-Browne, 1893) include descriptions of the geology of the Ely district. The stratigraphy of the Jurassic and Cretaceous rocks of the area are referred to in the stratigraphical memoirs *The Jurassic rocks of Britain* (Woodward, Vol V, 1895) and *The Cretaceous rocks of Britain* (Jukes-Browne, Vol 1, 1900; Vol II, 1903). Details of the water supply from underground sources in the district are given in the Geological Survey water supply memoirs for Cambridgeshire, Huntingdonshire and Rutland (Whitaker, 1922), Norfolk (Whitaker, 1921) and Suffolk (Whitaker, 1906). The borehole data in these memoirs was updated and revised in Wartime Pamphlet No. 20, Part II of which describes the water supply of the Ely Sheet from underground sources (Woodland, 1943).

A gravity survey of the district was carried out, as part of a regional study, in 1955 and published at a scale of a quarter- inch to one mile on Gravity Survey Overlay Sheet 16 in 1959. An aeromagnetic survey of the district is included in Sheet 2 (published 1965 at a scale 1 to 625 000) of the Aeromagnetic Map of Great Britain and Northern Ireland.

The memoir has been written and compiled by Dr Gallois and has been edited by Dr R. A. B. Bazley and Mr W. B. Evans. Specialist contributions on the palaeontology of the Jurassic by Dr H. C. Ivimey-Cook and Dr B. M. Cox, and of the Cretaceous by Mr A. A. Morter, have been incorporated in the account. Mr Seale has supplied descriptions of the soils and land use of the district. Drs D. E. Butler and D. E. White (Silurian–Devonian) and Mr C. J. Wood (Chalk) have provided palaeontological details of selected formations and have given stratigraphical advice. Mr F. G. Dimes has provided details of the building stones at Ely.

In poorly exposed areas of soft rock such as the Ely district the information obtained from boreholes and temporary sections is especially valuable. Only one deep borehole has been drilled within the district, for oil-shale exploration in 1921 at Methwold Common, but deep boreholes close to its boundaries, notably at Soham (Sheet 188) and Lakenheath (Sheet 174), have provided much information that enables the succession of the concealed strata beneath the Ely district to be predicted with confidence. The Geological Survey is indebted to the Superior Oil (UK) Ltd, for donating the cores of the Lakenheath Borehole for stratigraphical studies and for permission to publish the results described here.

The assistance is also gratefully acknowledged of Dr C. L. Forbes, Curator of the Sedgwick Museum at Cambridge, who has made available specimens and unpublished descriptions of temporary sections in the solid formations in the district.

F. G. Larminie, OBE
Director

British Geological Survey
Keyworth
Nottingham NG12 5GG
6 October 1987

CHAPTER 1

Introduction

GEOGRAPHICAL SETTING

The area described in this memoir is low-lying, sparsely populated, arable land lying to the north, west and east of the cathedral city of Ely, and includes parts of northern Cambridgeshire, south-western Norfolk and north-eastern Suffolk (Figure 1). It lies in the south-eastern part of Fenland and can be divided into two distinct types of land: Jurassic, Cretaceous and Pleistocene deposits that form low 'islands' of moderately well drained, populated ground; and an artificially drained, very sparsely populated plain of Recent deposits (reclaimed fens) that surrounds these 'islands'. More than 80 per cent of the district lies within Fenland: it includes almost the whole of the South Level and much of the Middle Level. In the following account the term 'district' is used to describe the area included in the 1:50 000 geological map, and 'region' to describe the north-western part of East Anglia.

In medieval and earlier times, when most of the fens were still undrained, settlements were restricted to the 'islands'. This pattern of settlement is still reflected in the present-day distribution of population, and some of the Fenland islands have probably been continuously inhabited since Neolithic times. The larger islands near the landward edge of Fenland, such as that on which Ely stands, have a complex settlement history because they were the first substantial areas of dry land that successive waves of sea-borne invaders encountered when they sailed up the Fenland waterways. Evidence of the various early medieval invasions remains in the place names of the district. Ely itself reputedly owes its origin as a centre of worship and trade to the foundation of a minster there, in a prominent position overlooking the River Great Ouse, in 673 AD by Etheldreda, daughter of a king of the East Angles. Church records chronicle later invasions by the Danes, Vikings and Saxons. By the time of the Domesday Survey all the major population centres in the district were well established.

Ely (population c. 11 000) is the major settlement. Much of the remaining population is concentrated in the large villages of Littleport, Southery, Sutton, Wimblington and Downham and the smaller villages of Coveney, Witcham, Witchford, Mepal, Pymore, Manea, Wentworth and Stuntney. The southern suburbs of March (population c. 15 000) lie in the north-western corner of the district.

The lowest point within the Ely district is probably about 1.5 m below Ordnance Datum in the fens near Shippea Hill, and the highest (about 26 m above OD) is in the southern part of Ely (Figure 1). Large parts of the Fenland area lie below sea-level due to peat wastage and most of the remaining Fenland area, with the exception of flood-banks, old river and tidal creek courses (roddons), and an area of young silt and clay around Welney, is less than 1 m above OD.

The district is drained by the Great Ouse and its tributaries by means of a complex network of drains. In pre-Roman times the western part of the district, now part of the Middle Level drainage system, drained westwards into the Ouse, which at that time followed a course downstream from St Ives that ran west of the March-Chatteris 'island' (Figure 1). The eastern part of the district was drained by the rivers Cam, Lark and Little Ouse, and these probably converged near Welney to form the Croft River, which ran to join the Ouse in a common outfall into The Wash at Wisbech. The Romans diverted the Ouse from near St Ives to Stretham, around the southern edge of the 'island' of Ely, and laid the foundations for the present drainage pattern. The combined waters of the Great Ouse, Cam, Lark and Little Ouse were subsequently diverted at Littleport to flow via Southery and Downham Market to The Wash at King's Lynn (Frontispiece). Reclamation of the peat lands of the southern fens began in earnest with the construction of the Old and New Bedford rivers in the 17th century under the guidance of the celebrated Dutch engineer Vermuyden. Reclamation continued, with a number of setbacks, until completed during the Second World War.

The district is almost entirely agricultural and has always been so. The only mineral worked at the present time is gravel for aggregate, although chalk, brick clay (Ampthill and Kimmeridge clays), phosphate (from the Woburn Sands, Gault and Cambridge Greensand), sand and gravel, and peat have been worked on a small scale in the past.

GEOLOGICAL SEQUENCE

The rock succession proved in the Ely district is summarised on the inside front cover: the geological setting is shown in Figure 2. Formations below the Ampthill Clay do not crop out in the district, and only the Oxford Clay and West Walton Beds have been proved in boreholes. However, Palaeozoic and Triassic rocks and a relatively complete Jurassic sequence have been proved in boreholes close to the margins of the district at Lakenheath [748 830] and Soham [5928 7448], and these rocks are likely to underlie the whole district. Predicted thicknesses of the concealed formations are given on the inside front cover: for completeness, lithological descriptions and thicknesses of the Palaeozoic and Triassic rocks proved in the BGS cored borehole at Soham, near the southern edge of the district, are included. Summaries of the sequences proved in the Lakenheath and Soham boreholes are given in Table 1.

GEOLOGICAL HISTORY

Little is known concerning the geological history of East Anglia in pre-Mesozoic times but it seems reasonable to assume that the sequence of sedimentation and earth-movements is similar to that of adjacent regions. Folded Precambrian and/or Palaeozoic rocks underlie most of East Anglia and the East Midlands at relatively shallow depths. The oldest rocks proved in the vicinity of the Ely district are indurated mudstones of late Silurian or early Devonian age in

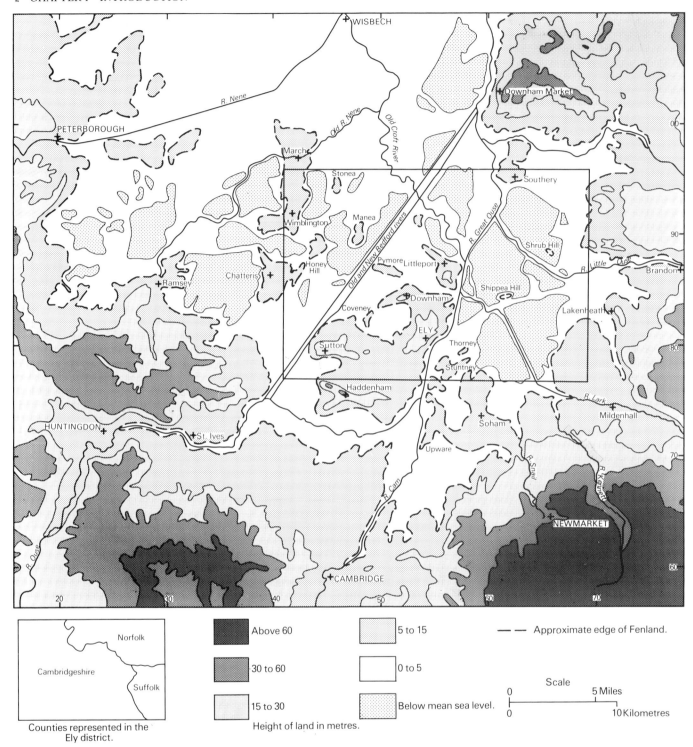

Figure 1 Geographical setting of the Ely district

the Lakenheath and Soham boreholes. These rocks appear to have been folded and lithified during the later stages of the Caledonian earth-movements to form a stable massif, part of the London Platform, which gave rise to islands throughout much of the Upper Palaeozoic and Mesozoic—the Wales-Brabant island in the Carboniferous, the Mercian Highlands in the Permian, the Anglo-Belgian island in the Jurassic and Lower Cretaceous (all from Wills, 1951).

During the Mesozoic the positions of the shorelines of the islands were more or less coincident at several different periods, suggesting that their boundaries were probably in part fault-controlled and that the faults were repeatedly active. In the mid and late Cretaceous the stable massif was finally overstepped by the successive transgressions of the Gault and Chalk.

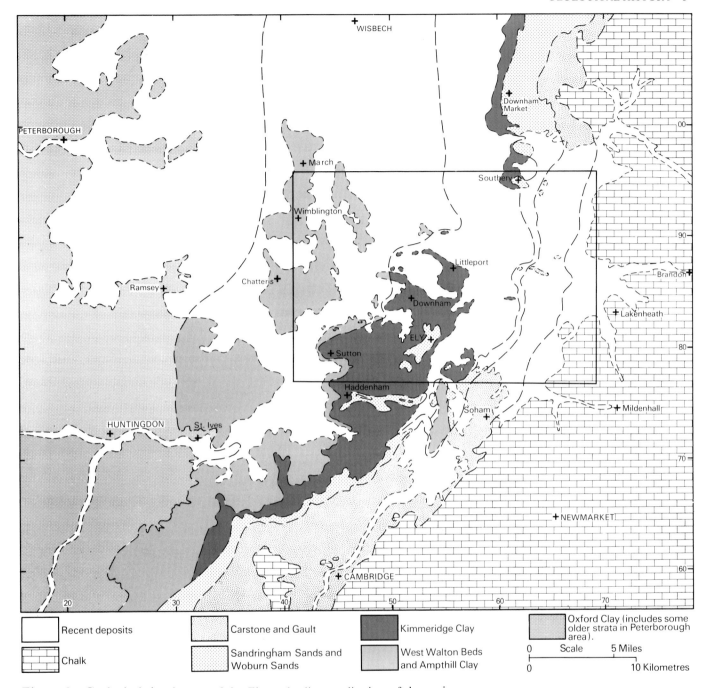

Figure 2 Geological sketch map of the Ely and adjacent districts of the region

Legend:
- Recent deposits
- Chalk
- Carstone and Gault
- Sandringham Sands and Woburn Sands
- Kimmeridge Clay
- West Walton Beds and Ampthill Clay
- Oxford Clay (includes some older strata in Peterborough area).

Scale: 0 — 5 Miles; 0 — 10 Kilometres

The oldest Mesozoic rocks proved in the close vicinity of the Ely district are red and green Triassic mudstones, sandstones and conglomerates that rest with markedly angular unconformity on folded Palaeozoic rocks. The Triassic rocks were deposited by flash floods, in ephemeral streams and playa lakes in a hot desert, and probably smoothed out and infilled an irregular topography of Palaeozoic rocks. The resulting low-relief land was then further eroded and transgressed by the warm, relatively shallow seas in which the Lias and late Jurassic rocks were deposited.

The Liassic sea moved across the Ely district in a southeasterly direction and eventually covered the whole of it.

Though only a part of the Lias is now preserved beneath most of the district, it is likely that the whole of it was formerly present but that much was removed by erosion prior to the deposition of Middle Jurassic sediments. The Lias of the Ely district is almost certainly composed of soft mudstones with a rich marine fauna including ammonites, bivalves, brachiopods, crinoids, foraminifera and marine reptiles, and is believed to have been deposited in relatively shallow water on a broad marine shelf under subtropical or tropical conditions. There is no direct evidence of the position of the Liassic shoreline in East Anglia, but the rapid thinning of the formation in the direction of the London Platform suggests

Table 1 Summary of strata penetrated in the Lakenheath and Soham boreholes

	Soham		Lakenheath	
	Thickness m	Depth m	Thickness m	Depth m
PLEISTOCENE				
River Terrace gravels	c. 3.0	c. 3.0	—	—
CRETACEOUS				
Chalk	absent		c.66.5	c. 66.5
Gault	c.19.5	22.45	c.19.8	c. 86.3
Carstone	c. 0.7	c. 23.2	c. 2.1	c. 88.4
Woburn Sands	c. 1.3	c. 24.5	presumed absent	
JURASSIC				
Ampthill Clay	absent		c. 3.4	c. 91.8
West Walton Beds	c.10.9	35.36	c. 4.1	95.86
Oxford Clay	41.53	76.89	c.37.0	c.132.9
Kellaways Beds	0.47	77.36	c. 2.1	c.135.0
Cornbrash	1.13	78.49		
Upper Estuarine 'Series'	1.54	80.03	presumed absent	
Lower Estuarine 'Series'	c.10.8	c. 90.8		
Lower Lias	c.42.3	133.07	c.47.6	182.6
TRIAS				
sandstones, marls and pebble beds	29.69	162.76	12.42	195.02
SILURIAN–DEVONIAN				
indurated mudstones	79.40	242.16	25.35	220.37
Final depth		242.16		220.37

that an island, probably heavily vegetated, lay not too far distant from the south-eastern corner of the Ely district and probably supplied some of the sediment for the mudstones.

Uplift towards the end of Liassic times produced erosion that removed the younger Liassic sediments, and caused the London Platform to become rejuvenated and to contribute to the coarser clastic sediments of the Middle Jurassic Estuarine 'Series'. These predominantly sandy deposits were laid down in shallow water on a broad coastal plain traversed by numerous small rivers and creeks draining through brackish lagoons and swamps. This coastal plain was inundated by the sea from time to time with the result that the shelly and sandy marine limestones of the Upper Estuarine 'Series' and Cornbrash were laid down in clear, shallow water.

Progressive deepening of the sea, combined with an increase in the supply of siliciclastic material, partly from the London Platform, gave rise to the Kellaways Beds and then to the Oxford Clay, West Walton Beds and Ampthill and Kimmeridge clays. The presence of mudstones rich in calcareous debris derived from pelagic marine fauna and flora shows that at times the Upper Jurassic sea was clear and relatively free from land-derived mud. Periodically, notably during the Oxfordian, the sea was locally sufficiently warm, clear and shallow to support coral reefs, and the currents sufficiently strong close to the London Platform to produce shoals of calcareous oolite. This type of sequence is well displayed in the Upware area, immediately to the south of the Ely district, and probably extends beneath the south-east corner of the district.

The Jurassic rocks that crop out in the district contain a rich marine fauna. The Ampthill and Kimmeridge clays are characterised by ammonites and bivalves, with subordinate numbers of brachiopods, echinoids, crinoids, serpulids, foraminifera, ostracods, crustacea, fish and aquatic reptiles. The nearby land that occupied much of the central part of East Anglia was probably covered by tropical forests and cut by sluggish rivers. Some of the clastic material in the Jurassic clays appears to have been derived from the weathering products of Palaeozic mudstones that cropped out on this massif.

On land the dinosaurs were the dominant group of vertebrates at that time; flying reptiles, the pterosaurs, and probably the first birds were also present, together with a few small mammals and amphibians. The land probably carried a vigorous and diverse plant life including ferns, horsetails, cycads, conifers and ginkgos. The vertebrate fauna of the sea was dominated by the marine reptiles, the ichthyosaurs and plesiosaurs, and by fish.

Earth-movements, including gentle folding, affected the district towards the end of the Jurassic period causing erosion of the Upper Jurassic sediments and rejuvenation of the London Platform. As a result, the Ely district was transformed into an area of predominantly sandy sedimentation with evidence of deposition in shallow water under the action of strong tidal currents. These earth-movements were the first of a number of discrete phases that began shortly after the deposition of the Kimmeridge Clay and continued into the Cretaceous, the last phase ending a little before the deposition of the Woburn Sands. They were the continental reflections of widespread earth-movements (the 'Cimmerian' orogeny) associated with the separation of the European and North American continental plates and the opening of the North Atlantic.

As a result of these earth-movements, the youngest Jurassic rocks in the Ely district, the Roxham Beds, are sands with a marine fauna of bivalves and ammonites, interbedded with plant- rich clays; both lithologies seem likely to have been deposited in shallow water close to land. The Roxham Beds overstep the Kimmeridge Clay in a south-easterly direction, and a pebble bed at their base contains phosphatised fragments of ammonites derived from the highest part of the Kimmeridge Clay. In the southern part of the district the Roxham Beds have themselves been removed by erosion and fossiliferous debris from them and the Jurassic clays is incorporated in the basal pebble bed of the Lower Greensand.

The present-day relationship of the Lower Greensand to the Jurassic rocks in the district shows that folding, uplift and extensive erosion occurred during the early Cretaceous. Around the edges of the London Platform the Lower Greensand consists of relatively coarse sands and pebbly sands that rest on a planed surface cut in Jurassic and older rocks. The Lower Greensand of the Ely district is sparsely fossiliferous, a feature that probably reflects its unsuitability as a medium for fossil preservation rather than an originally sparse fauna. Elsewhere in southern England the Lower Greensand is characterised by a rich fauna dominated by bivalves but with brachiopods, cephalopods, sponges, polyzoa, echinoids and gastropods locally common; the relative abundance of plant debris and bones of terrestrial vertebrates in some horizons suggests the proximity of land.

The London Platform appears to have been rapidly eroded during the period of deposition of the Lower Greensand and became so degraded that it was subsequently submerged and was overstepped by the mudstones of the Gault. The contrast between the turbulent conditions of the Lower Greensand and those in which the Gault deposited is reflected in the contrasting faunas. The generally thick-shelled and coarsely-ribbed molluscs of the Lower Greensand are replaced by mostly thin, smooth-shelled forms in the Lower Gault. Benthonic faunas became increasingly more sparse with time in the Gault so that the highest beds of the Gault and the succeeding Chalk are composed largely of the calcareous parts of coccoliths and other pelagic algae.

Although only the lowest part of the Chalk is now exposed in the Ely district it is likely that the formation was formerly fully represented. The purity of the soft limestones of the Chalk suggests that the nearest source of siliciclastic material was some considerable distance away and that the London Platform had little or no influence on their sedimentation. Both the fauna and lithology of the Chalk suggest deposition in warm, clear water at depths which were probably greater than were present at any time during the late Jurassic or early Cretaceous.

The marine faunas of the Upper Jurassic and Cretaceous are essentially similar in character with bivalves and ammonites the dominant invertebrates and reptiles the dominant vertebrates. On land the dinosaurs reached their maximum size in the Cretaceous, but the flying reptiles had reached their acme in the Jurassic and were less important than the birds in the Cretaceous. Small placental mammals may also have been present. A marked change occurred in the land-flora during the Cretaceous. Angiosperms, the ancestors of modern flowering plants, appeared and became a substantial part of the flora by the mid Cretaceous, subsequently dominating the remaining conifers, cycads and ferns.

Earth-movements at the end of the Cretaceous caused world-wide uplift and erosion and large areas of what had been the chalk-sea were converted to land or to shallow-water lakes, estuaries and shallow-water marine shelves on which were deposited muddy and sandy sediments. This change was also marked by the reduction or extinction of several groups of animals. Ammonites and belemnites became extinct, brachiopods had already become much reduced and world-wide changes occurred in the composition of the planktonic foraminifera. By this time the dinosaurs, plesiosaurs and pterosaurs had also become extinct and the age of the 'ruling reptiles' was ended.

No evidence of Tertiary deposition has been recorded in the district and it seems likely that for most of the period it was an area of erosion. During the latter part of the period, during the Pliocene, deposition may have been resumed, but any sediments are likely to have been composed of loose sands and pebbly sands that have been rapidly removed by erosion during the Quaternary.

The Quaternary history of the area is complex, but represented by only a fragmentary record. At least one ice sheet passed over the district, removed much of any high ground that was present and infilled the lower ground and hollows with boulder clay and gravelly sand. Subsequently, periglacial climates prevailed for long periods and at other times fluviatile and marine gravels were deposited under temperate conditions.

Since the time of the maximum extent of the last ice-sheet (about 18 000 years ago) sea-level has risen by more than 100 m and has caused the inundation of the Fenland part of the district. Erosion products washed down from the higher ground have accumulated in this area as a complex sequence of brackish, marine and fresh-water deposits.

PREVIOUS WORK

The earliest references to the geology and physical geography of the Ely district occur in Anglo-Saxon and medieval charters and chronicles. Among these the *Liber Eliensis* is of particular interest since it traces the early history of Ely and, in doing so, describes its setting and agricultural economy. Observations on the geology occur in works relating to the drainage of Fenland and those describing the agriculture of the counties of Cambridgeshire and Norfolk. Skertchly (1877, pp.307 – 314) has provided a comprehensive bibliography of these early publications together with a list of works published between 1702 and 1876 relating to the geology of the Fenland part of the district. A geological bibliography for the Cambridgeshire part of the district covering the period 1794 to 1879 is given in Penning and Jukes-Browne (1881, pp.171 – 180).

The first published geological map of the district is that incorporated in William Smith's maps of England and Wales

(1815) and the counties of Cambridgeshire (1819) and Norfolk (1819). Few works have been specifically devoted to the district but the descriptions by Hailstone (1816), Lunn (1819), Bonney (1875), Reed (1897), Rastall (1910) and Brighton (1938) of the geology of Cambridgeshire and those by Woodward (1833), Rose (1835–1836), Trimmer (1846), Gunn (1866), Harmer (1877) and Larwood and Funnell (1961) of the geology of Norfolk have provided general accounts that include parts of it. The stratigraphical works by Roberts (1892) on the Upper Jurassic rocks, Woodward (1895) on the Jurassic rocks and Jukes-Browne (1900–1903) on the Cretaceous rocks describe particular aspects of the local geology.

The most famous exposure in the Ely district, the huge borrow-pit at Roslyn Hole, Ely, was the cause of much geological controversy in Victorian times and was visited by numerous field parties. Fisher (1868), Seeley (1865a, 1865b, 1868), Bonney (1872a, 1872b) and Skertchly (1877) all contributed to the debate as to whether a large exposure of Cretaceous rocks in the pit was *in situ* or part of a large erratic mass within the boulder clay. These visits provided a great deal of incidental information on the Kimmeridge Clay that occupied the bulk of the pit.

One-inch-to-one-mile Geological Survey maps of the district were published in 1881–1883 (part of Sheet 51) and 1886 (part of Sheet 65). The surveys were made by Bennett, Bristow, Cameron, Hawkins, Jukes-Browne, Penning, Reid, Skertchly and Woodward, and the accompanying memoirs were written and compiled by Whitaker, Woodward and Jukes-Browne (1891) and Whitaker, Skertchly and Jukes-Browne (1893). The geology of the Fenland part of the district had previously been described by Skertchly (1877) in his classic Fenland memoir, and the whole of its natural history by Miller and Skertchly (1878).

More recent work in the district has concentrated on obtaining a better understanding of the relationships of particular local deposits to those of adjacent areas. The works of Pringle (1923) on the concealed Mesozoic rocks, Gallois and Cox (1976, 1977) on the Upper Jurassic clays, Casey (1961a, 1961b, 1967), Rastall (1919, 1925) and Schwartzacher (1953) on the Sandringham Sands and the Lower Greensand, Baden Powell (1934, 1948), Woodland (1970), Sparks and West (1965), Sparks and others (1972), Straw (1960, 1979) and West and Donner (1956) on the Pleistocene deposits and Fowler (1932, 1947), Godwin (1938a, 1938b, 1940), Godwin and Clifford (1938), Seale (1974, 1975) and Willis (1961) on the Recent deposits have enabled much of the geology of the district to be placed within its regional setting. The results of these, and works of more local interest, are discussed in the main text.

Some of the limited amount of mineral working in the district has been referred to in more general works. Grove (1976) has described the coprolite workings in the Lower Greensand and Gault, and a number of regional accounts refer to clay, chalk, sand and gravel workings. Whitaker (1906, 1921, 1922) and Woodland (1943) have described the underground water supply of the district. Seale (1975) has described the soils of the district in a comprehensive memoir and maps, and has related them to the geology, landscape and land use.

The Fenland part of the district was until relatively recently a wild and mysterious place, and much has been written about its reclamation. Parts of it have been inhabited almost continuously since Neolithic times and the history of its drainage works probably extends back to that period. Phillips (1970) and Darby (1940a, 1940b) have summarised the Roman and Medieval works respectively and Godwin (1978) has provided an eloquent summary of the archaeology and natural history of the area.

CHAPTER 2

Precambrian and Palaeozoic

Precambrian and Palaeozoic rocks have been proved in boreholes at relatively shallow depths throughout the east Midlands and the north-western part of East Anglia. No economically interesting mineral has yet been proved to be present in these basement rocks and the presence of only a thin Mesozoic cover makes it unlikely that there is any major accumulation of hydrocarbons in the region. In consequence, no deep borehole has been drilled within the Ely district and few have been drilled in the adjacent area.

Seismic investigations by Bullard and others (1940) suggested that the Palaeozoic floor was less than 150 m below ground level beneath parts of the Cambridge district and, as a result, the Geological Survey drilled continuously cored boreholes through the Mesozoic cover into Palaeozoic rocks at Cambridge [4316 5949] and close to the southern boundary of the Ely district at Soham [5928 7448], in 1952–1953 and 1955 respectively (Holmes *in* Worssam and Taylor, 1969). A seismic refraction survey carried out by Dr J.D. Cornwell at Padnal Fen [575 828] in 1979 indicated Mesozoic rocks overlying probable Palaeozoic rocks at about 180 m below OD

Hydrocarbon exploration boreholes have been drilled to the north and east of the Ely district at Glinton, Spalding, Wisbech, Wiggenhall, Lakenheath, Breckles, Rocklands and Great Ellingham, in areas where the preserved Mesozoic succession is thicker than at Ely. Boreholes drilled for stratigraphical, site investigation or water supply purposes have also penetrated the basement in the Huntingdon, Ramsey and Bury St Edmunds districts (Figure 3).

The Ely district is situated on the northern flank of the London Platform and is probably underlain by Precambrian or Palaeozoic rocks at depths of about 150 to 250 m below OD; the depth increases northwards across the district (Figure 3).

STRUCTURE

The structure of the pre-Mesozoic rocks of the district is not known but the borehole data combined with aeromagnetic and gravity surveys enables their probable structure in the region to be determined in general terms.

The aeromagnetic map of the Ely and adjacent districts (Figure 4) shows the departure of the local geomagnetic field from the background field, a linear geomagnetic field compiled from a network of magnetic observations made throughout Great Britain. The magnetic anomalies of the region are mostly broad and of small amplitude except in the area between Peterborough and Wisbech, in the north-western part of Fenland. Kent (1968) and Linsser (1968) both assumed the positive magnetic anomalies in the region to be due to basic igneous rocks within the crystalline basement. In the areas of negative anomaly, in the southern part of the region (Figure 4), it has been suggested (Linsser,

1968) that these basic rocks are at depths of more than 6 km; he estimated them to be at depths of 3 to 6 km in the positive anomaly area. With the exception of a diorite proved in the Warboys Borehole, Cambridgeshire (IGS, 1966), no magnetic rock has yet been recorded in the Palaeozoic or Mesozoic of the region. If magnetic rocks are generally absent at this stratigraphical level this could imply that 3 km to more than 6 km of Palaeozoic/Mesozoic sediments overlie the Precambrian. The borehole evidence and the gravity data (see below) do not, however, support this interpretation.

The Bouguer gravity anomaly map (Figure 5) has been computed for an idealised geoid based on the 1967 International Gravity Formula. The data is corrected for elevation so that the resulting anomalies can reliably be attributed to variations in the density, and hence the gravitational attraction, of the rocks in the outer layers of the earth.

The gradients of the Bouguer anomalies in the Ely and adjacent districts are low. Chroston and Sola (1975) and Howell (*in* Cox and others, in press) have analysed the gravity anomaly pattern for the northern part of East Anglia and have both concluded that it largely reflects the depth to the Precambrian basement. In addition, Howell suggested that crustal thinning has a marked effect in the north Norfolk coast area where it gives rise to a positive gradient as the coast is approached. Linsser (1968) and Chroston and Sola (1975) drew attention to the apparent disparity between the aeromagnetic and gravity data for large parts of East Anglia. The basic rocks postulated by Linsser to be responsible for the positive magnetic anomalies would be dense rocks that would give rise to positive gravity anomalies if they were at shallow depths. Figure 4 shows a pattern of mostly positive magnetic anomalies in the northern part of the region with negative anomalies in the south. The gravity map (Figure 5) shows the opposite trend. To explain this Linsser suggested that the magnetic anomalies might be due to dense basic rocks that had subsided into the crust to leave hollows that had subsequently become infilled with thick sedimentary sequences. However, the mechanism of such a process remains questionable and there is no supporting borehole evidence.

A more satisfactory explanation is that the gravity anomalies broadly reflect the depth to the density-contrast surface that probably occurs at the junction of the Precambrian and the overlying rocks. The magnetic anomalies are probably related to magnetic rocks within the Precambrian and the depth to these, relative to the density-contrast surface, may vary considerably within the region.

Similar trends occur in the aeromagnetic and gravity maps over small areas and one of these, a NW–SE orientated trend running across the Ely district, may indicate an area where the upper surface of the magnetic basement is close to that of the Precambrian. The regional trend suggests that the depth to the magnetic basement probably increases southwards from about 3 to 6 km, with the Ely district in the

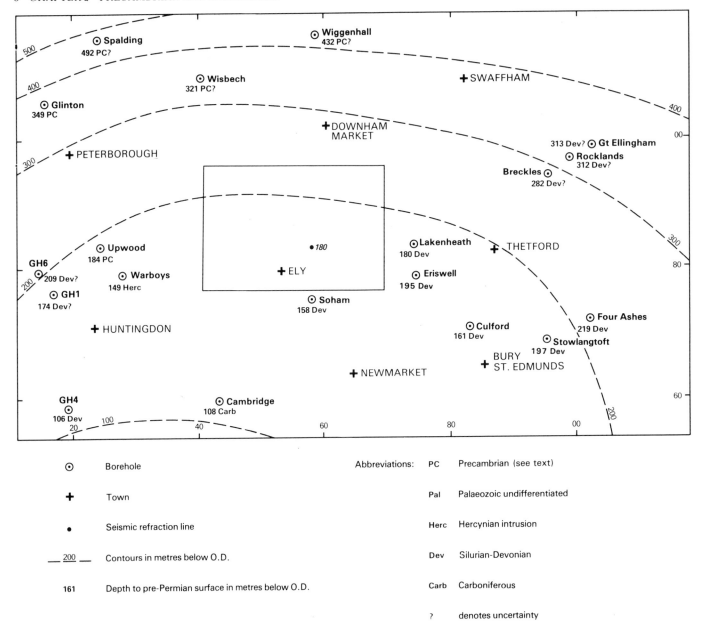

⊙	Borehole	
✛	Town	
•	Seismic refraction line	
─ 200 ─	Contours in metres below O.D.	
161	Depth to pre-Permian surface in metres below O.D.	

Abbreviations:

PC	Precambrian (see text)
Pal	Palaeozoic undifferentiated
Herc	Hercynian intrusion
Dev	Silurian-Devonian
Carb	Carboniferous
?	denotes uncertainty

Figure 3   Contours on the upper surface of the pre-Permian rocks of the region

medium to deeper part of this range. In the south-western part of the district a local positive anomaly may reflect igneous rocks within the Palaeozoic. Steep magnetic gradients at the western end of this anomaly are probably caused by the Hercynian diorite proved in the Warboys Borehole.

The depth to Precambrian rocks cannot be determined from the gravity anomaly map without a knowledge of the lithologies and densities of these basement rocks. Estimates by Chroston and Sola (1975) and Howell (in press) of 0.5 to 1.0 km to the Precambrian in the north-eastern part of the region seem likely to be correct. The presence of volcaniclastic rocks of presumed Precambrian age at 349 m below OD at Glinton, Northants (Kent, 1962) and 721 m below OD at North Creake, Norfolk (Kent, 1947) have been interpreted as high points on the Precambrian upper surface separated by a thick Palaeozoic sequence beneath Fenland.

Data from a seismic refraction line run from The Wash to Lowestoft have been interpreted by Chroston (1985) as indicative of crystalline basement rocks, either Precambrian metamorphics or later igneous intrusions, at depths varying from about 800 m in the west to 2 km in the east. In the eastern part of the region, and including much of the Ely district, the depth to the Precambrian seems likely to be everywhere greater than 1 km. Comparison of the gravity anomaly map (Figure 5) with that showing the depth to the pre-Permian surface (Figure 3) suggests that variations in the thickness of the Mesozoic rocks have little influence on the regional pattern of gravity anomalies.

The structure of the older Palaeozoic rocks of the region is clearly complex. Where observed in boreholes, such as those at Culford, Eriswell, Four Ashes, Great Ellingham, Lakenheath, Soham and Stowlangtoft, these rocks are

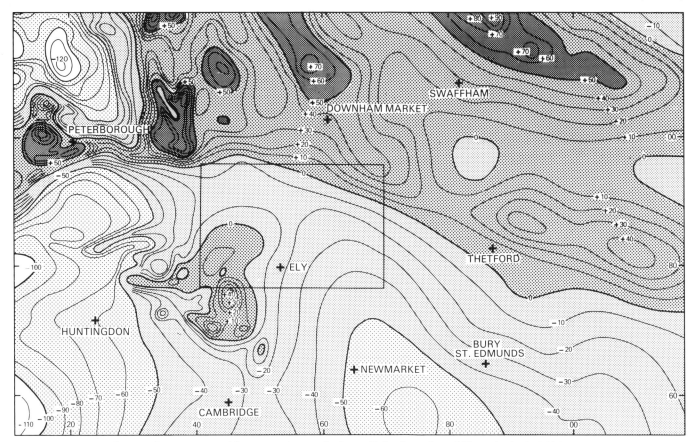

Contour values represent total force magnetic anomalies in nanoTeslas above a linear field equation for the British Isles which implies a regional increase in total force of 2.1728 nanoTeslas per km. northwards and 0.259 nanoTeslas per km. westwards (National Grid directions), and a datum value of 47033 nanoTeslas at the grid reference origin for epoch 1955.5.

Contour interval 10 nanoTeslas with thicker lines at 50 nanoTeslas

−200 to −100 nanoTeslas

−100 to −50 nanoTeslas

−50 to 0 nanoTeslas

0 to +50 nanoTeslas

+50 to +100 nanoTeslas

Figure 4   Magnetic anomaly map of the region

lithologically similar to one another and have yielded faunas or floras indicative of a late Silurian to mid Devonian age. Dips have been more than 20° and commonly more than 45°. The almost horizontal Palaozoic rocks proved at Lakenheath are the single exception yet recorded. The indurated nature, degree of fracturing and poorly developed cleavage of these beds shows that they have been compressed, but not intensely so. Their consistent age over a large area of East Anglia suggests the presence of either an extremely thick sedimentary sequence (more than 15 000 m) or a moderately thick sequence whose subcrop beneath the Mesozoic is repeated by folding and/or faulting.

PRECAMBRIAN

In the area immediately to the north and west of the Ely district the rocks of presumed Precambrian age encountered

in boreholes have included rhyolitic tuffs at Glinton [TL 1500 0528] (Kent, 1962) and Upwood [TL 2493 8304] (Horton and others, 1974), metamorphosed sandstones at Spalding [TF 2434 1478] and Wisbech [TF 4066 0843] and metamorphosed mudstones at Wiggenhall [TF 5914 1537]. The age of these last three occurrences have always been in considerable doubt and they probably form part of a thick sequence of early Palaeozoic sediments. The Glinton and Upwood occurrences were thought to extend westwards, beneath a relatively thin Mesozoic cover, to the Precambrian outcrops of Charnwood Forest in Leicestershire. However, recent analyses suggest that the basement rocks in these and other boreholes in the East Midlands form part of an early Palaeozoic (probably Ordovician) volcaniclastic complex (T. C. Pharoah, personal communication).

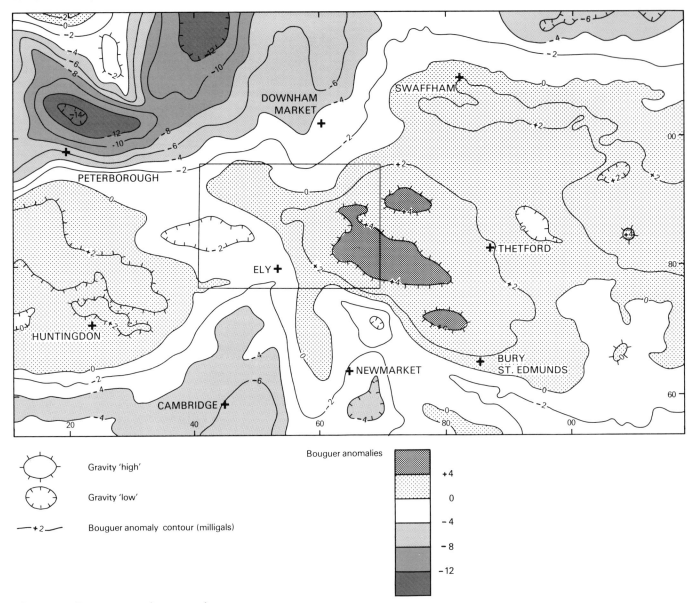

Figure 5   Bouguer gravity anomaly map

PALAEOZOIC

Silurian – Devonian

The rocks at Upwood, Wisbech and Wiggenhall that have been assigned to the Precambrian are siliciclastic and volcaniclastic sediments that have been sufficiently thermally metamorphosed to suggest that they are significantly older than the fossiliferous Palaeozoic rocks proved in the region. No Cambrian or Ordovician rock has been proved but, by analogy with rocks of this age in the Midlands and Wales, they would be expected to be folded but of low metamorphic grade. The basement rocks proved in boreholes in the area immediately south and east of the Ely district consist mostly of steeply dipping, indurated, poorly cleaved, marine mudstones of late Silurian or early Devonian age: the district itself is probably largely underlain by similar sediments. The limited evidence available suggests that these Palaeozoic

Plate 1   Sedimentary structures in the Silurian – Devonian of the Soham Borehole. All × 1

a. Small slump folds in silty mudstone with secondary quartz vein filling possible dewatering structure (BDK 2577)
b. Finely laminated siltstone and muddy siltstone showing small scour hollow (top); cross-lamination (top left); possible hummocky cross-lamination (bottom); *Chondrites*-type burrows and possible escape burrow (arrowed) (BW 5649)
c. Climbing ripples (ripple-drift bedding) in siltstone; resting with scoured contact on finely laminated siltstones with a few small burrows and a larger (possible escape) burrow at the contact (BW 5634)
d. Pale silt bands and laminae within a predominantly mudstone sequence, showing load structures disturbed by burrowing (BW 5812)
e. Thinly interbedded and interlaminated mudstone and siltstone with numerous burrows (BDK 2588)
f. Graded bedding in thin fining-upward units in mudstone and muddy siltstone (BW 5797)

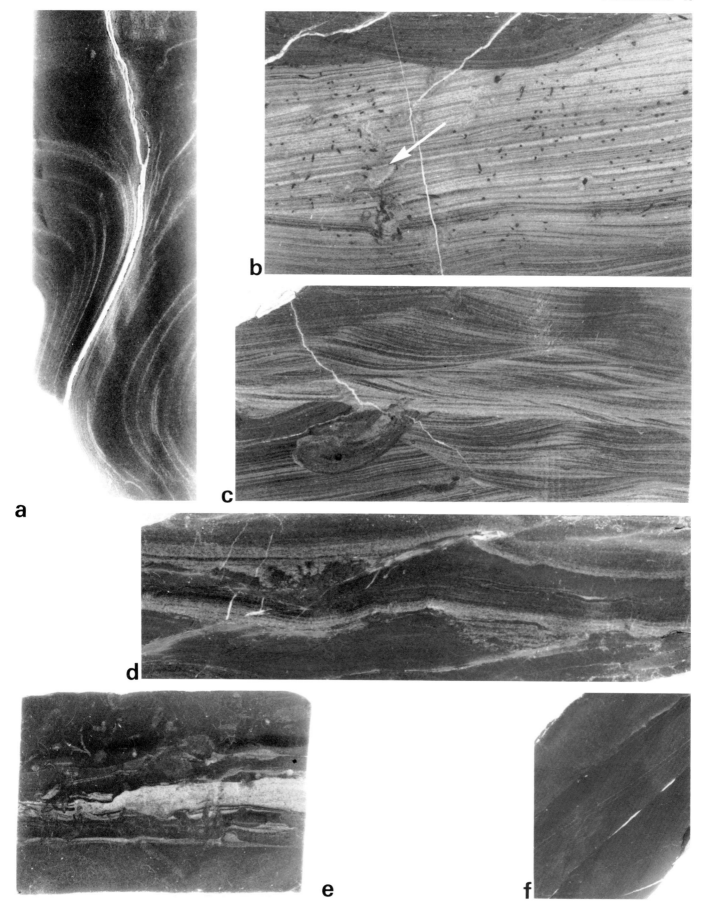

rocks form part of a thick sequence that has been extensively folded and that the interface with the Precambrian is now affected by this folding.

The Lakenheath and Soham boreholes proved rhythmically bedded mudstones that contain faunal and sedimentary evidence indicative of deposition on an offshore marine shelf (Plate 1). In the Lakenheath Borehole, the Palaeozoic rocks are almost horizontal grey mudstones with thin siltstone bands, purple and red-stained in their upper part, that yielded a marine fauna of poorly preserved bivalves, brachiopods, gastropods and ostracods. Indurated red and green (upper part) and grey mudstones with thin siltstone bands dipping at 45° to 85° were proved in the Soham Borehole (Holmes *in* Worssam and Taylor, 1969, p.8). The poorly preserved marine fauna at Soham includes bivalves, brachiopods and a tentaculitid (Stubblefield *in* Mortimore and Chaloner, 1972). A restricted assemblage of microfossils and chitinozoa is present at Soham and suggests an early Devonian age (Chaloner and Richardson *in* House and others,1977).

Since 1972, the internationally agreed position for the Silurian–Devonian boundary has been the base of a particular graptolite zone as recognised at a type section at Klonk, Czechoslovakia. Graptolites have not been recorded at this stratigraphical level in Britain where marine faunas are known only from a few isolated sections in boreholes. At outcrop, in the Welsh borderlands, this part of the succession is represented by brackish and freshwater sediments of the 'Downtonian' and 'Dittonian' stages. In that area, the Silurian–Devonian boundary is believed to fall within the 'Downtonian' Stage (House and others, 1977, fig. 7). The sequences proved in the Lakenheath and Soham boreholes have some faunal and floral affinities with parts of the 'Downtonian', but cannot yet be correlated with the Klonk sequence. In our present state of knowledge therefore, they are considered to be late Silurian and/or early Devonian in age.

In mid or late Devonian times, during the later phases of the Caledonian earth-movements, the Palaeozoic mudstones of the region were folded and lithified to form the London Platform. The age of the folding cannot be accurately determined but it clearly post-dates the sediments at Culford, Great Ellingham, Four Ashes, Lakenheath and Soham (all late Silurian to mid Devonian in age) and pre-dates horizontally bedded soft marls and shelly limestones of late Devonian (Frasnian) age that have been proved elsewhere in East Anglia (Gallois *in* Cox and others, in press; Butler, 1981).

Carboniferous

Over 100 m of Carboniferous Limestone was proved in the Cambridge Borehole (Holmes *in* Worssam and Taylor, 1969, p.7), but no strata of this age have been recorded in the adjacent districts and their subcrop appears to be restricted to a small area around Cambridge.

DETAILS

Soham Borehole

The following sequence was cored between 163.4 and 242.2 m in the Geological Survey borehole at Soham:

	Thickness m	Depth m
Mudstone, silty in part; dark purplish red with orange-red (Triassic staining) patches and with open joints infilled with dolomite in top part; becoming pale greenish grey and dark grey with dull purplish and brownish red staining with depth; graded bedding present throughout, mostly in units 1 to 10 cm thick; sole structures and bioturbation, including *Chondrites*, common; thin interbeds of micaceous, finely planar, cross- and ripple-laminated siltstones at several levels; sparsely shelly throughout but with some bedding planes crowded with poorly preserved bivalves and ostracods; shelly limestone composed largely of gastropods at one level	79.40	242.2

The dip in the cores is steep, mostly 45° to 55° in the upper part increasing to 65° to 85° in the lower, but is not inverted. A dipmeter survey showed that the direction of dip between 196 and 215 m was relatively steady and was on average 054°. A marine fauna, dominated by thin-shelled bivalves and brachiopods, is present and was attributed to the Devonian by Stubblefield (in Mortimer and Chaloner, 1972). Mortimer (1967) recorded a sparse pollen flora, including *Nematothallus* ? and cf. *Spongiophyton* indicative of a possible Lower Devonian age. The fauna has been re-examined by Dr D.E. Butler (unpublished Ph.D., London 1977) who recorded *Lingula punctata* Hall, *Lingula sp.*, *Lindstroemella sp.*, *Protochonetes ?*, *Retichonetes ?*, rhynchonellaceans indet., '*Murchisonia*' *sp.*, *Conocardium sp.*, *Cypricardella* cf. *gregaria* (Hall & Whitfield), *Goniophora sp.*, *Limoptera* aff. *normanniana* (d'Orbigny), *Modiomorpha sp.*, '*Nuculana*' *sp.*, *Nuculites sp.*, *N.* cf. *truncatus* (Steininger), *Palaeoneilo sp.*, *Palaeosolen sp.*, *Parallelodon* aff. *hamiltoniae* (Hall & Whitfield), *Prothyris* aff. *plicata* Spriestersbach, *Sanguinolites ? sp.*, an orthocone, *Dicricoconus sp.*, *Tentaculites sp.*, a trilobite fragment and the ostracods *Londinia sp.*, *Macrypsilon ? salterianum* (Jones) and smooth forms. Dr Butler noted that the fauna was partly Silurian and partly Devonian in aspect and concluded that it was late Silurian or early Devonian (late 'Downtonian' or early 'Dittonian') in age.

Lakenheath Borehole

A lithologically similar sequence to that in the Soham Borehole was proved beneath the Triassic in the Lakenheath Borehole. Two cores were taken (at 195.02 to 199.95 m and at 217.32 to 220.37 m), the higher one in mottled purplish red and pale green mudstones with thin siltstone interbeds, and the lower in similar mudstones and siltstones, but grey and unstained. These beds at Lakenheath are horizontal and less fractured and stained than the Silurian–Devonian at Soham but are otherwise very similar in lithology and in their included sedimentary structures. Graded bedding, planar and cross-laminated (including ripple-drift) bedding, convolute bedding (due to either loading or slumping) and bioturbation are present in both sequences (Plate 1). The Lakenheath cores have yielded a fauna that is similar in character and probably in age to that from Soham. The smaller faunal list from Lakenheath is probably a reflection of the limited amount of core available for study. Dr Butler (1977) recorded bivalves, including *Nuculites ? antiqua* (J. de C. Sowerby) and *Prosocoelus ? sp.*, brachiopods, and ostracods including *Londinia sp.* and cf. *Frostiella groevalliana* Martinsson, and also thought this fauna was probably referable to the late Silurian–early Devonian. Dr D.E. White, having compared the faunas of both boreholes with those of the Little Missenden Borehole, Buckinghamshire, and the Stonehouse Formation of Arisaig, Nova Scotia, has concluded that they are likely to be of mid- to late 'Downtonian' age.   RWG

CHAPTER 3

Triassic

Triassic rocks are probably present everywhere beneath the Ely district. Their regional distribution suggests that they thicken northwards in an irregular manner, but that they are unlikely to exceed 50 m in thickness anywhere within the district (Figure 6).

In the Soham Borehole, about 30 m of soft brown and red, partially dolomitised sandstone and red silty mudstone rest on a weathered and eroded surface of Palaeozoic rocks. The lowest 10 m contain angular, (probably thermally shattered), fragments of Palaeozoic mudstone and sandstone at several levels and these beds are presumed to have been deposited in poorly sorted fans and screes. In the Lakenheath Borehole, about 12 m of similar dolomitised sandstone and red mudstone rest with a planar and apparently unweathered junction on the Palaeozoic. The sequence there is finer grained overall than at Soham but both sequences contain evidence, such as dolomite/anhydrite concretions, breccias, mud-pellet conglomerates, insolation structures and desiccation cracks, indicative of deposition in a hot desert subjected to flash floods. No faunal or floral data was recorded at this

stratigraphical level in either borehole, but the distribution and age of the Triassic in East Anglia and the southern North Sea suggests that the sequences, such as those of the Ely district, that are close to the London Platform and the southern limit of preserved Triassic rocks, are late Triassic in age.

DETAILS

Soham Borehole

The following Triassic sequence was proved between 133.1 and 163.4 m in the Soham Borehole and is probably typical of that beneath much of the Ely district:

	Thickness m	Depth m
Siltstone, greenish grey and reddish brown interbedded with micaceous fine-grained sandstone; patchy dolomitic cement and nodular dolomite concretions (? replacing anhydrite) at several levels; mud-pellet conglomerates at two levels; passing down into	6.5	139.6

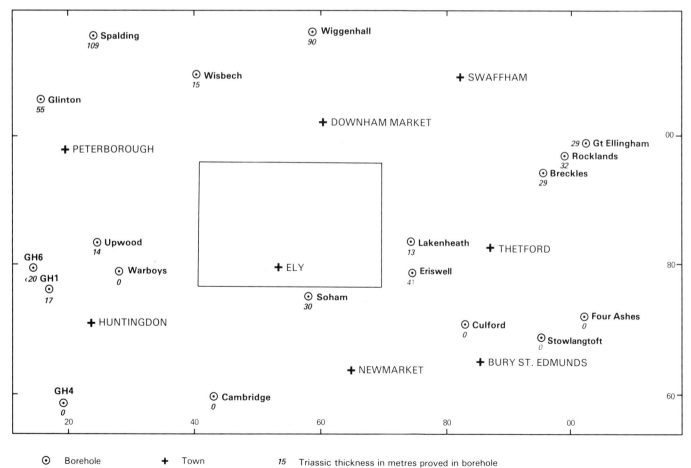

Figure 6 Thicknesses of the Triassic rocks of the region

	Thickness m	Depth m
Mudstone, silty and muddy siltstone; dull red with some pale green banding and mottling; several thin sandstone interbeds; traces of horizontal lamination; sharp base	6.7	146.3
Sandstone, coarse-grained, pebbly with minor thin beds of red silty mudstone; pebbles mostly Triassic sandstone and mudstone and well-rounded vein quartz; coarse conglomerate at base includes angular small boulders of Triassic mudstone resting on a very irregular surface	10.7	157.0
Sandstone, mottled dull red and greyish green; fine-grained becoming pebbly with depth; dolomitic cement at several levels; pebble bed at base and sharp junction with	5.7	162.7
Mudstone, silty, mottled red and green, crowded with angular fragments of dark purplish red mudstone from underlying bed; very irregular junction with Palaezoic mudstones	0.7	163.4

The sequence is crudely rhythmic, each fining-upward rhythm composed of a basal pebble bed or conglomerate that passes up into pebbly sandstone and then into sandstone and, in one example, into marl. Each rhythm or partial rhythm probably represents the torrential deposits of a flash flood followed by quieter fluviatile or lacustrine deposition.

Lakenheath Borehole

A thinner and generally more fine-grained sequence than that of the Soham Borehole was proved between 182.6 and 195.0 m in the Lakenheath Borehole. There, soft, mostly fine-grained, micaceous sandstones with patchy dolomitic cement are interbedded with thin beds of red and green, mottled mudstones. These latter rest with a planar junction on red-stained, but otherwise unweathered, horizontal Palaeozoic mudstones, and appear to be fluviatile and playa-lake deposits laid down in relatively quiet conditions.   RWG

CHAPTER 4

Jurassic

Representatives of the Lower, Middle and Upper Jurassic[1] are present beneath much of the Ely and adjacent districts, but are attenuated in comparison with other parts of southern England. The Jurassic sequence is particularly thin adjacent to the London Platform, but thickens rapidly westwards towards the Midlands and northwards into the southern North Sea (Figure 7). Evidence of erosion, condensed deposition and near-shore environments occur at many levels. Three major and a large number of minor breaks in sedimentation are present. More substantial unconformities occur at the base of the Lias (here of

1 To enable direct comparisons to be made with earlier geological literature and the published geological maps, the term Upper Jurassic is used in this memoir sensu Arkell (1947), and includes the deposits of the Callovian, Oxfordian, Kimmeridgian and Volgian stages.

Pliensbachian age), at the base of the Middle Jurassic sequence (Bajocian), and at the base of the Sandringham Sands (Volgian).

The thickness of the Jurassic rocks now preserved in the Fenland region has been much reduced by erosion during the Quaternary and possibly during the Tertiary. Close to the London Platform, erosion also occurred during the early Cretaceous and removed part or all of the Jurassic. It is possible from the known preserved thicknesses of the various formations, together with a knowledge of their regional thickness trends, to estimate the probable total thickness of the Jurassic rocks of the region prior to Tertiary and Quaternary erosion. No estimate can be made of the amount of sediment removed by erosion surfaces within the Jurassic, or

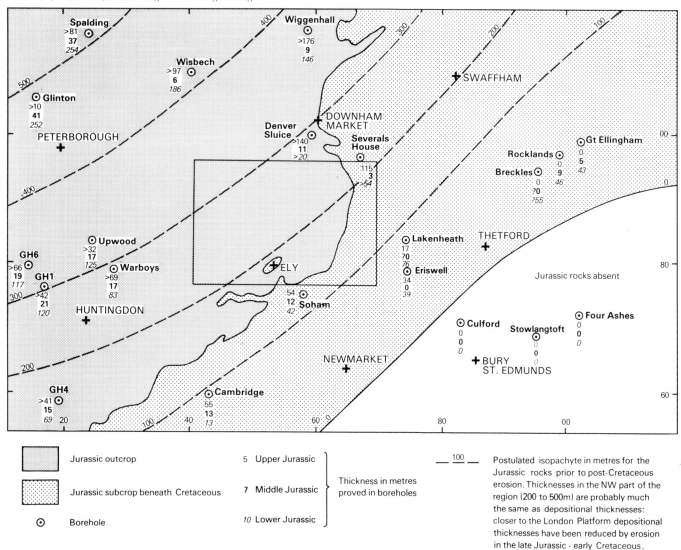

Figure 7 Thicknesses of the Jurassic rocks of the region

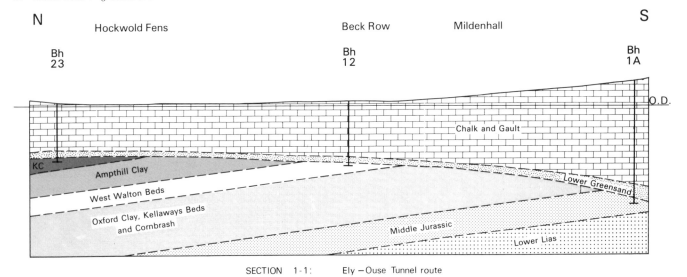

N   Hockwold Fens   Beck Row   Mildenhall   S

SECTION 1-1:   Ely—Ouse Tunnel route

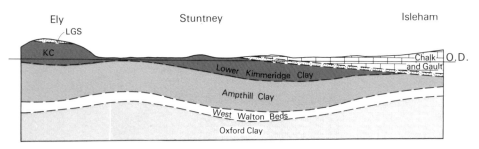

NW   SE

SECTION 2-2 : Ely to Isleham

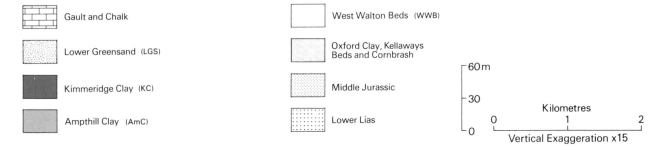

                                                            ESE
WNW

SECTION 3-3: Haddenham to Fordham (after Holmes and others, 1965)

Gault and Chalk

Lower Greensand (LGS)

Kimmeridge Clay (KC)

Ampthill Clay (AmC)

West Walton Beds (WWB)

Oxford Clay, Kellaways
Beds and Cornbrash

Middle Jurassic

Lower Lias

60 m

30

0

Kilometres

0   1   2

Vertical Exaggeration x15

Figure 8a Overstep of Cretaceous rocks across folded Jurassic rocks in the south-eastern part of the Ely district (opposite)

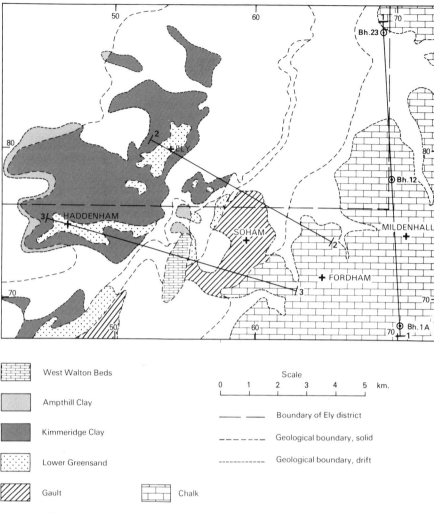

Figure 8b Geological sketch map of south-east Fenland showing lines of section referred to in Figure 8a

West Walton Beds

Ampthill Clay

Kimmeridge Clay

Lower Greensand

Gault

Chalk

Scale

0   1   2   3   4   5   km.

——— ——— Boundary of Ely district

– – – – – – Geological boundary, solid

---------- Geological boundary, drift

Recent deposits left blank : Pleistocene deposits omitted for clarity.

by Cretaceous erosion around the edge of the London Platform, because there are too little data on which to base regional trends. Despite these limitations, the Jurassic rocks of the region can be shown to thicken north-westwards: their total depositional thicknesses appear to have ranged from about 100 m to more than 300 m in the Ely district (Figure 7).

Only the upper part of the Jurassic crops out in the district but the remainder of the succession is known from nearby boreholes, notably those at Soham, Lakenheath and Severals House [6921 9639], Methwold. The Lower and Middle Jurassic sequences are comparable in lithology and fauna with those at outcrop in the East Midlands where they have been described in the various Geological Survey memoirs and reviewed by Woodward (1895), Arkell (1933) and Hallam, Sylvester-Bradley, Torrens and Callomon (*in* Sylvester-Bradley and Ford, 1968). The nomenclature and zonation used to describe the formations of this age in the present account are those published for the East Midlands outcrop.

Between Oxfordshire and the River Humber, the Upper Jurassic is composed almost entirely of soft mudstones. These rocks weather easily and give rise to a broad belt of low-lying ground that is largely obscured by Quaternary

deposits. William Smith (1819–1820), in his geological maps of the counties of Bedfordshire, Cambridgeshire, Norfolk and Suffolk, recognised two Upper Jurassic formations, the Oxford Clay and the overlying Oaktree Clay (subsequently re-named the Kimmeridge Clay). Seeley (1861) suggested that, in Bedfordshire and Cambridgeshire, the Oxford Clay and the Kimmeridge Clay were separated by an additional clay formation which he later (Seeley, 1869) termed the Ampthill Clay taking the name from sections at Ampthill, Bedfordshire. He correlated this clay with the mixed argillaceous, arenaceous and calcareous beds grouped elsewhere in England under the name Corallian Beds, and also (Seeley, 1861) drew attention to the presence of silty and calcareous mudstones and muddy limestones lying between the Oxford Clay and the Ampthill Clay in the area around Elsworth, Cambridgeshire. These lime-rich beds are laterally variable in lithology and have been given a number of local names, including Elsworth Rock (Seeley, 1861), Elsworth Series (Wedd, 1898), Oakley Beds (Arkell, 1933), St Ives Rock (Seeley, 1861) and Upware Coral Rag/Corallian Oolite (Blake and Hudleston, 1877), some of which may be synonyms.

Until recently, the term Elsworth Rock Group (Edmonds and Dinham, 1965) was used for all the silty and calcareous beds between the Oxford and Ampthill clays, but these beds have never been fully exposed and there has been no type section or precise definition of their stratigraphical limits. The boundaries of this unit have now been lithologically defined on the basis of a type section in a continuously cored borehole at West Walton, Norfolk, and the name West Walton Beds has been proposed for them (Gallois and Cox, 1977).

On his geological map of Norfolk (1819), William Smith showed a sandy formation overlying the Kimmeridge Clay. The bulk of this formation is made up of poorly exposed, soft sands that were later termed the Sandringham Sands (Whitaker and Jukes-Browne, 1899). These sands are sparsely fossiliferous and were believed to belong with the Lower Cretaceous because they rest unconformably on the Kimmeridge Clay. However, in 1961, sections through the basal beds of the Sandringham Sands, exposed in drainage works at West Dereham (Sheet 159), yielded ammonites indicative of a late Jurassic age (Casey 1961b, 1973). Work in the Sandringham Sands outcrop between West Dereham and Heacham, Norfolk, subsequently showed that they could be divided into four members, the Roxham Beds and the Runcton Beds (Jurassic) and the Mintlyn Beds and the Leziate Beds (Cretaceous) (Casey and Gallois, 1973). Only the oldest of these, the Roxham Beds, is present in the Ely district due to the southerly overstep of a younger Cretaceous formation (the Carstone) across the Sandringham Sands.

A complete Upper Jurassic sequence from the Cornbrash to the Roxham Beds is probably present everywhere beneath the Ely district but only the Ampthill Clay, Kimmeridge Clay and Roxham Beds crop out in the district. The West Walton Beds, and possibly the upper part of the Oxford Clay, have a subcrop beneath a thick sequence of till in a glacially-filled valley at March, but there are insufficient borehole data available to plot the formation boundaries accurately beneath the drift.

The structure of the Jurassic rocks is simple; over most of the district they dip very gently eastwards to south-eastwards at 1° or less. Low-amplitude folds affect the Jurassic rocks of the southern part of the Ely district where they can be demonstrated to be truncated by the Lower Greensand (Figure 8, Section 2.2). In the Cambridge district Holmes (in Worssam and Taylor, 1969, Plate III) postulated the presence of a number of NNE–SSW trending synclinal and anticlinal axes affecting the Jurassic; but even within these folds the steepest proven dips do not exceed 1°. One of them, the Upware Anticline, continues into the Ely district as a very gentle flexure running along the line of the valley of the Great Ouse, and causes beds lower in the Kimmeridge Clay than would have been anticipated from a steady eastwards dip to crop out on the eastern sides of the 'islands' of Ely and Littleport (Figure 8, Section 3.3).

A slight discordance of dip can also be demonstrated between the Jurassic and Cretaceous rocks in the eastern part of the Ely district, between Hockwold and Mildenhall, where the Lower Greensand has been proved in boreholes to rest on progressively older Jurassic formations in a southerly direction, as the London Platform is approached (Figure 8, Section 1.1).

LOWER JURASSIC

The Lower Jurassic of the East Midlands can be divided into three predominantly argillaceous formations, the Lower Lias (mudstones and calcareous mudstones), Middle Lias (silty, sandy and ferruginous mudstones) and the Upper Lias (bituminous and calcareous mudstones). Thin representatives of all three formations have been proved in boreholes in or adjacent to the Ely district. Lower Lias is likely to be present everywhere beneath the district, Middle Lias is probably preserved beneath the northern part of the district, and some Upper Lias is probably present in the north-western part.

The Lower Lias thins south-eastwards across the district due to three factors, all of which are related to the proximity of the London Platform. Firstly, beds belonging to progressively younger zones of the Lias overstep onto the Trias in a south-easterly direction (Donovan and others, 1979, fig. 4); secondly, the Middle Jurassic disconformably oversteps the Lower and Middle Lias in the same direction; and thirdly, minor erosion surfaces within the Lias become more pronounced as the London Platform is approached. Lower Lias was proved in the Soham Borehole where it was directly overlain by Middle Jurassic. The geophysical evidence (electric logs) from the Lakenheath Borehole indicates that the Lower Lias there is unconformably overlain by Upper Jurassic. Thin representatives of the Middle and Upper Lias and the Middle Lias respectively were present in boreholes close to the present district at Denver Sluice [TF 5911 0106] (Gallois, 1979) and at Severals House (Pringle, 1923).

DETAILS

No borehole has penetrated the Lias in the district but, taken together, the sequences proved in the Soham, Lakenheath and Severals House boreholes encompass most of the stratigraphical range of the Lias that is likely to be present.

Soham Borehole

The best documented section through the lower part of the Lias in the region is that proved in the Soham Borehole where the following sequence was recorded between 90.8 and 133.07 m:

	Thickness m	Depth m
Tragophylloceras ibex Zone		
Mudstone, deeply weathered passing down into unweathered pale and medium grey silty mudstone with rare ammonites and pyritised trails; thin beds of tabular clay ironstone, pyritic shelly mudstone, shelly siltstone and shelly limestone with oysters and other bivalves, rhynchonellids and belemnites occur at several levels	c.8.3	99.1
Siltstone interbedded with silty mudstone; medium and pale grey becoming brown and purplish red in lower part; calcite and limonite cements at several levels; shelly in part with common bivalves and rare ammonites; dense ferruginous oyster-rich limestone lens at one level; very shelly in lowest part with much pyritized shell brash and common ammonites, belemnites, oysters and rhynchonellids; bioturbated junction with	c.29.3	128.4

	Thickness m	Depth m

Uptonia jamesoni Zone

Mudstone, pale grey, silty interbedded with brashy, weakly cemented very shelly mudstone with common bivalves, brachiopods and ammonites; passing down into — c.3.4 131.83

Siltstone, brownish grey, oolitic (limonite) passing down into densely oolitic mudstone; sparsely shelly but with bivalves, brachiopods and ammonites locally common; pebble bed at base, with pebbles of Triassic mudstone, sandstone and dolomite, resting on irregular burrowed surface of Triassic — 1.24 133.07

RWG

Uptonia jamesoni Zone faunas occur between 128.40 and 132.71 m, and there is no reason to doubt that the underlying 0.36 m of strata, down to the base of the Lias, can be included in this zone. Forms of *Uptonia bronni* (Roemer) are common between 128.40 and 130.30 m, many being close ribbed. This species is regarded by Donovan (in Dean and others, 1961) as characterising the higher parts of the *jamesoni* Zone. A large oxycone at 130.90 m and *Apoderoceras sp.* at 132.71 m indicate lower parts of the zone. The non-ammonite fauna is locally rich in brachiopods especially in the lower part. Taxa present include *Cincta numismalis* (Lamarck), *C. nummosa* S.S. Buckman, *Cirpa fronto* Quenstedt, *Lobothyris radstockiensis* (Davidson), *Spiriferina* cf. *tumida* (von Buch) and abundant *Zeilleria waterhousei* (Davidson). Bivalves present include *Chlamys substriata* (Roemer), *Oxytoma inequivalve* (J. Sowerby), *Parainoceramus ventricosus* (J. de C. Sowerby), *Pholadomya ambigua* (J. Sowerby), *Plicatula sp.* and *Pseudolimea sp.*

Between the highest *Uptonia sp.* at 128.40 m and lowest *Acanthopleuroceras valdani* at 126.87 m lie 1.53 m of beds with crinoid fragments, *Cincta pernummus* S. S. Buckman, bivalves and *Passaloteuthis*. These beds may belong with the *Tropidoceras masseanum* Subzone of the *Tragophylloceras ibex* Zone, but this cannot be proved in the absence of the characteristic ammonites.

The remainder of the Lower Lias is attributed to the *ibex* Zone. The *Acanthopleuroceras valdani* Subzone is well proved between 126.18 m and 126.87 m with abundant specimens of the subzonal index together with *Tragophylloceras ibex* (Quenstedt), brachiopods, bivalves and belemnites. Above this bed occurs a thick sequence of mudstone with only scarce ammonites including *Tragophylloceras sp.* at 107.95 m and *Liparoceras* at 119.48 m, below the lowest *Beaniceras sp.* at 105.54 m. These beds contain a rich crinoid, brachiopod and bivalve fauna and are provisionally attributed to the *valdani* Subzone.

Beaniceras was found infrequently between 94.56 m and 105.54 m together with one *Tragophylloceras sp.* at 104.88 m, and probably indicates the *Beaniceras luridum* Subzone of the *ibex* Zone, though *Beaniceras* first appears in the *valdani* Subzone. The poorly fossiliferous beds between the highest ammonite and the top of the Lower Lias can also be attributed to this subzone. The *luridum* Subzone yielded a rich fauna of crinoids, brachiopods, bivalves and other taxa including *Ditrupa quinquisulcata* (Münster), *Balanocrinus subteroides* (Quenstedt), *Cincta sp.*, *Lobothyris sp.*, *Cuneirhynchia sp.*, *Piarorhynchia* cf. *juvenis* (Quenstedt), *Tetrarhynchia?*, *Antiquilima sp.*, *Astarte sp.*, *Camptonectes mundus* Melville, *Cardinia attenuata* (Stutchbury), *Cucullaea sp.*, *Dacryomya gavei* (Cox), *Eopecten* cf. *tumidus* (Zeiten), *Eotrapezium cucullatum* (Goldfuss), *Goniomya hybrida* (Münster), *Grammatodon insons* (Melville), *Gryphaea gigantea* J. de C. Sowerby, *G. suilla* Chapuis & Dewalque, *Hippopodium sp.*, *Laevitrigonia? troedssoni* Melville, *Liostrea sp.*, *Lithophaga?*, *Lucina limbata* Terquem & Piette, *Mactromya arenacea* (Terquem), *Modiolus numismalis?* Oppel, *M. scalprum* J. Sowerby, *Myoconcha decorata* (Münster), *Nuculana (Rollieria) bronni* (Andler), *Oxytoma inequivalve*, *Palaeoneilo galatea* (d'Orbigny), *Parainoceramus ventricosus*, *Pholadomya*

ambigua (J. Sowerby), *Pinna sp.*, *Plagiostoma sp.*, *Platymya?*, *Pleuromya costata* (Young & Bird), *Plicatula calloptychus* (Deslongchamps), *Plicatula spinosa* J. Sowerby, *Pronoella intermedia* (Moore), *Protocardia truncata* (J. de C. Sowerby), *Pseudolimea acuticosta* (Münster), *Pseudopecten acuticosta* (Lamarck), *Pseudopis sp.*, *Ryderia doris* (d'Orbigny), *Tutcheria cingulata* (Goldfuss), *T.* cf. *richardsoni* Cox; *Dentalium sp.*, *Amberleya (Eucyclus) subimbricata* (d'Orbigny); *Passaloteuthis* and *Pseudohastites*.   HCIC

Lakenheath Borehole

The thickness of the Lower Lias in the Lakenheath Borehole (c. 47.6 m) is comparable to that at Soham and probably covers a similar stratigraphical range. Only part of the Lower Lias (from 146.30 to 173.70 m) was cored. This part of the sequence can be matched lithologically and faunally with part of that from Soham. The ammonite fauna indicates that parts of the *jamesoni* and *ibex* zones are present. The Lower Lias at Lakenheath is probably unconformably overlain by Upper Jurassic.   RWG

The presence of the *jamesoni* Zone was proved between the lowest recorded ammonite, a *Polymorphites quadratus* (Quenstedt) at 172.60 m indicating the *Platypleuroceras brevispina/Polymorphites polymorphus* Subzone, and a *Dayiceras dayiceroides* (Mouterde) at 158.10 m. These two lower subzones cannot be reliably separated. They also contain *Platypleuroceras* aff. *brevispina* (J. de C. Sowerby) at 170.61 m and *Platypleuroceras* up to 166.45 m together with *Tragophylloceras*, early *Uptonia* and an apoderoceratid. It is probable that all the underlying core, and possibly the remainder of the Lias down to its base, can be attributed to the *jamesoni* Zone. The beds between 166.45 and 158.10 m contain *Uptonia* without *Platypleuroceras* or *Polymorphites* and are attributed to the *jamesoni* Subzone.

The non-ammonite fauna of the zone is dominated by bivalves including *Astarte*, *Camptonectes*, *Cardinia*, *Grammatodon*, *Laevitrigonia?*, *Myoconcha*, *Palaeoneilo*, *Palaeonucula*, *Plagiostoma*, *Plicatula*, *Protocardia*, *Pseudolimea*, *Ryderia* and *Tutcheria*. Crinoid columnals, *Cincta*, *Lobothyris*, proceritheid gastropods, *Hastites*, *Passaloteuthis* and *Pseudohastites* also occur.

The *ibex* Zone was proved between 149.42 and 157.96 m. *Tropidoceras* was recorded at several levels between 157.73 and 157.96 m and may indicate the presence of the *masseanum* Subzone as this genus appears at the base of the *ibex* Zone but persists into the *valdani* and *luridum* subzones. The presence of the *valdani* Subzone is indicated by the presence of *Acanthopleuroceras* between 157.17 and 153.16 m whilst *A. valdani* (d'Orbigny) occurs between 157.0 and 156.3 m, *Liparoceras sp.* at 156.28 m and *Tragophylloceras sp.* at 154.10 m. A specimen of *Beaniceras?* at 149.42 m may indicate the presence of the overlying *luridum* Subzone.

The non-ammonite macrofauna of the *ibex* Zone is comparable to that of the *jamesoni* Zone and is likewise dominated by bivalves which occur in even greater variety although brachiopods are less numerous. *Cardinia attenuata* (Stutchbury) is abundant at several levels. Hastitid belemnites become more common in the higher beds.   HCIC

No palaeontological data were recorded from the uncored (c. 135.0 to 146.3 m) part of the Lias in the borehole but comparison of the geophysical logs, in particular the resistivity log, with those from continuously cored boreholes in Oxfordshire (Horton and Poole, 1977) suggests that only the *ibex* Zone is present. A marked resistivity peak at 145.1 m at Lakenheath may be the '85ft Marker' of Horton and Poole (1977) which is approximately coincident with the base of the *luridum* Subzone.   RWG

Severals House Borehole

The highest 43.3 m of the Lower Lias were partially cored in the Severals House Borehole (Pringle, 1923). The sequence is composed of mudstones, silty mudstones and thin muddy limestones from which a sparse fauna of ammonites, bivalves, echinoderms and gastropods was collected. Poorly preserved, smooth ammonites that occur at 197.2 and 198.12 m may be *Gamellaroceras sp.* indicative of the lower part of the *jamesoni* Zone. Pringle (1923) recorded an ammonite identified by S.S. Buckman as *Aegoceras* cf. *latecosta* (J. de C. Sowerby) at 161.5 m; this was re-identified by Spath (MS 1941) as '*Beaniceras costatum* S.S. Buckman transitional to *Androgynoceras maculatum* var. *intermedia* Spath'. The specimen is now lost but the identification provides evidence of the *Prodactylioceras davoei* Zone. The intervening beds contain *Piarorhynchia sp.*, *Astarte sp.*, *Camptonectes mundus*, *Grammatodon sp.*, *Oxytoma sp.*, *Palaeoneilo galatea*, *Pseudopecten sp.*, *Ryderia sp.* and echinoderm fragments.

The Lower Lias at Severals House is overlain by Middle Lias, probably with minor unconformity. Pringle (1923, p.131) allocated 11.74 m (from 149.19 to 160.93 m) of micaceous sandy mudstone and rusty brown, muddy limestone to the formation. The fauna includes crinoid debris, belemnite fragments and amaltheid ammonites including *Amaltheus subnodosus* (Young & Bird) at 157.4 m. This last named indicates the presence of the *Amaltheus subnodosus* Subzone of the *Amaltheus margaritatus* Zone.   HCIC

MIDDLE JURASSIC

The Middle Jurassic sequence is condensed everywhere adjacent to the western edge of the London Platform and is composed of brackish and freshwater sediments similar to those of the Lower and Upper Estuarine 'Series' of the east Midlands. Representatives of these beds probably occur beneath the whole of the Ely district with the exception of a small area in the south east (Figure 7). Where proved in boreholes, they range from 5 to 15 m in thickness and probably thin irregularly south-eastwards as they approach the London Platform. Predominantly fresh and brackish water sequences of sands, limestones and clays (Lower and Upper Estuarine 'Series') were proved in the Soham and Severals House boreholes; these 'Series' are assigned to the Bajocian and Bathonian stages respectively. The distribution of oolitic limestone facies (Lincolnshire Limestone, Great Oolite Limestone etc.,) in the Middle Jurassic in the East Midlands suggests that their most south easterly limit lies close to, but outside, the north-western part of the Ely district.

DETAILS

Soham Borehole

The most complete Middle Jurassic sequence recorded to date in the region is that in the Soham Borehole where the following succession was proved between 78.49 and c.90.8 m:

	Thickness m	Depth m
Upper Estuarine 'Series'		
Mudstone, finely laminated pale greenish grey; preserved as lithorelics between cracks and burrows filled with Cornbrash	0.02	78.51
Limestone, dense, silty, pale slightly greenish and pinkish greys; becoming less calcareous with depth and passing down into bioturbated siltstone; becoming shelly with depth with oysters and bivalves	0.33	78.84
Oyster lumachelle composed almost entirely of *Liostrea hebridica* (Forbes); passing down into very shelly, silty mudstone and siltstone with bivalves and gastropods; very irregular bioturbated junction with bed below	1.19	80.03

Lower Estuarine 'Series'		
Siltstone and very fine-grained sandstone; pale grey and brownish grey; carbonised vertical plant rootlets ubiquitous in upper part; coalified wood fragments abundant at several levels; becoming silty and clayey with depth with sphaerosiderite pellets giving rise to purplish red and yellowish brown staining; bioturbated junction with Lias	c.10.8	90.8

The fauna of the beds between 78.49 and 80.03 m, which includes *Anisocardia*, *Corbulomima*, *Liostrea*, *Modiolus*, *Placunopsis*, *Praemytilus*, *Pseudolimea* and *Vaugonia*, is typical of the Upper Estuarine 'Series' of the Northamptonshire/Lincolnshire outcrop. The low diversity and composition of the fauna suggests a marine environment with brackish influence (cf. Hudson, 1963; 1980). The beds below the erosion surface at 80.03 m, barren except for rootlets and wood fragments, are comparable with the freshwater parts of the Lower Estuarine 'Series' (Bajocian) and with parts of the Upper Estuarine 'Series' (Bathonian) and could belong to either. The abrupt nature of the erosion surface and the absence of fauna may indicate that they are Bajocian.   RWG

Severals House Borehole

The few specimens that have survived from the Middle Jurassic of the Severals House Borehole are similar in lithology and fauna to those from Soham, and the Soham sequence is, therefore, taken to be typical for the whole of the Ely district. In the Severals House Borehole, Pringle (1923, p.131) described about 3.0 m (from 146.2 to 149.2 m) of interbedded dark grey shelly clays and limestones which he allocated to the 'Forest Marble and Cornbrash'. The fauna, which includes echinoderm fragments, an ornithellid, *Entolium?*, *Gervillella sp.*, *Liostrea* cf. *hebridica* (Forbes), *Modiolus sp.* and *Placunopsis socialis* Morris & Lycett, is similar to that of the Blisworth Clay and Blisworth Limestone in the upper part of the Great Oolite 'Series'. The lithologies are compatible with this interpretation.   HCIC

UPPER JURASSIC

A complete Upper Jurassic sequence is present beneath the eastern part of the Ely district but is thin in comparison with that of the East Midlands and other parts of southern England. The older formations (Cornbrash, Kellaways Beds, Oxford Clay and West Walton Beds) are everywhere concealed and have only been proved in boreholes (Table 2). The younger formations (Ampthill Clay, Kimmeridge Clay and Roxham Beds) crop out in the western and central parts of the district and occupy more than two thirds of its total area. The current zonal scheme for this part of the sequence is given in Table 3.

CORNBRASH

Although some representative of the Cornbrash is likely to be present beneath much of the Ely district, it has been positively identified only in the Soham Borehole. The sequence there consists of 1.1 m of limestone and mudstone resting unconformably on a Middle Jurassic sequence, and is similar to that of the Northamptonshire outcrop where a thin, tough, grey limestone is overlain by a bed of ferruginous marl. It seems likely that these lithologies are present beneath much of the Ely district. Pringle (1923, p.137) allocated 0.6 m (from 146.2 to 146.8 m) of grey limestone in the Severals

House Borehole to the Cornbrash on the basis of lithology in
the absence of characteristic fossils. The only specimens of
the limestone to have survived have been classified as
Blisworth Limestone in the present work (see p.20). Any
representative of the Cornbrash that is present is likely to be
Upper Cornbrash as this has a much more extensive outcrop
and subcrop than the Lower Cornbrash. The geophysical
logs of the Lakenheath Borehole suggest that the Lower Lias
and Oxford Clay are separated by 2.1 m of limestone and
sand: this probably includes some representative of the Upper Cornbrash. RWG

DETAILS

Soham Borehole

The following succession was proved in the Soham Borehole between 77.36 and 78.49 m:

	Thickness m	Depth m
Upper Cornbrash		
Macrocephalites macrocephalus Zone (pars)		
Mudstone, greyish brown with burrowfill wisps of rotted limonitised shell debris; shelly in lower part with oysters, myids, pectinids, rare ammonites, serpulids and pyritised trails; resting on irregular bored surface	0.31	77.67
Limestone, densely cemented, sparsely shelly with *Entolium* and other bivalves, a solitary coral and shell debris; phosphatic pebble close to base; intensely bioturbated junction with Upper Estuarine 'Series'	0.82	78.49

No diagnostic fauna was recorded from the limestone, but the
overlying mudstone yielded a single *Macrocephalites* and indicates a
Callovian age. There is no evidence to suggest the presence of
Lower Cornbrash (Bathonian).

KELLAWAYS BEDS

Thin representatives of the Kellaways Beds were proved in
the Soham and Severals House boreholes and the formation
is probably widely represented in the Ely district. The best
documented section proved to date in the region is that in the
Soham Borehole where 0.4 m of silty and sandy Kellaways
Beds are present. The lithologies can be matched with those
of the outcrop along the western margin of Fenland. The
Kellaways Beds appear to be thin everywhere in the Ely
district. Pringle (1923) recorded about 1.5 m of clayey sand
and sandy marl at the base of the Oxford Clay in the Severals
House Borehole but did not consider this to represent the
Kellaways Beds. However, both the lithology and fauna of
specimens collected by Pringle can be matched with material
from the Kellaways Beds of the East Midlands outcrop. The
geophysical logs of the Lakenheath Borehole suggest that a
thin sandy and/or limey representative of the Kellaways
Beds is probably present there.

DETAILS

Soham Borehole

The following sequence was recorded in the Soham Borehole between 76.89 and 77.36 m:

	Thickness m	Depth m
Kellaways Rock		
Sigaloceras calloviense Zone (pars)		
Siltstone, greyish brown, weakly calcite-cemented; very shelly with common *Gryphaea (Bilobissa) dilobotes* Duff, belemnites and some ammonite fragments; passing down with bioturbation into	0.30	77.19

Table 2 The Upper Jurassic formations of the Ely district

Stage	Formation	Lithology	Thickness in Ely district
Volgian/ Portlandian	Sandringham Sands (Roxham Beds)	Fine-grained glauconitic and clayey sands	up to 6 m
Kimmeridgian	Kimmeridge Clay	Soft mudstones, calcareous mudstones and oil shales	up to 46 m
Oxfordian	Ampthill Clay	Soft mudstones and calcareous mudstones	10 to 50 m
	West Walton Beds	Silty mudstones, calcareous mudstones and muddy limestones	5 to 14 m*
	Oxford Clay	Soft mudstones, calcareous mudstones and bituminous mudstones	36 to 42 m*
Callovian	Kellaways Beds	Silty and sandy mudstones	up to 1 m*
	Cornbrash (Upper)	Limestone and mudstone	up to 2 m*

* Proved in boreholes only

Table 3 Zonal scheme for the Oxfordian, Kimmeridgian and Volgian/Portlandian sequences of the Ely district

Stage and substage			Zone	Formation (Ely district)
Volgian (pars)[1]	Portlandian (pars)[2]		Paracraspedites oppressus Titanites anguiformis Galbanites kerberus Galbanites okusensis Glaucolithites glaucolithus Progalbanites albani	Roxham Beds
Kimmeridgian	Kimmeridgian	Upper	Virgatopavlovia fittoni Pavlovia rotunda Pavlovia pallasioides Pectinatites pectinatus Pectinatites hudlestoni Pectinatites wheatleyensis Pectinatites scitulus Pectinatites elegans	Kimmeridge Clay
		Lower	Aulacostephanus autissiodorensis Aulacostephanus eudoxus Aulacostephanus mutabilis Rasenia cymodoce Pictonia baylei	
Oxfordian		Upper	Amoeboceras rosenkrantzi Amoeboceras regulare Amoeboceras serratum Amoeboceras glosense	Ampthill Clay
		Middle	Cardioceras tenuiserratum Cardioceras densiplicatum	West Walton Beds
		Lower	Cardioceras cordatum Quenstedtoceras mariae	Oxford Clay (pars)

1 Sensu Casey, 1967; 1973

2 Sensu Cope and others, 1980

∿∿∿ erosion surface

| | | | absent because of erosion

	Thickness m	Depth m
Kellaways Clay *Macrocephalites macrocephalus* Zone (pars) Mudstone, greyish brown, silty, pyritic; single large phosphatic nodule; barren except for rare bivalves; bioturbated junction with Cornbrash	0.17	77.36

The mudstone yielded no zonally diagnostic fossil but is assigned, by analogy with the East Midlands sequence, to the phosphatic facies of the Kellaways Clay (*macrocephalus* Zone, *M. kamptus* Subzone (Callomon, 1968)). The Kellaways Rock yielded fragmentary *Proplanulites* and *Sigaloceras*; the latter confirms the presence of the *calloviense* Zone.

Severals House Borehole

Specimens collected by Pringle from 146.15 m in the Severals House Borehole are typical Kellaway Beds lithologies and consist of bioturbated, very sandy, shelly mudstone with common oysters, *Meleagrinella* and *Oxytoma*, and a few kosmoceratid ammonite fragments. On the basis of Pringle's (1923) lithological descriptions the formation was probably encountered at 144.78 to 146.15 m in the borehole.

BMC, RWG

OXFORD CLAY

A complete Oxford Clay sequence is present at relatively shallow depths everywhere beneath the Ely district. Although the full thickness of the formation has been penetrated in boreholes in the adjacent districts at Denver Sluice, Severals House, Lakenheath and Soham, only one borehole in the district, that at Methwold Common [678 941], has reached the formation. Few samples of Oxford Clay have survived from the Methwold Common and Severals House boreholes and the formation was not cored in the Lakenheath Borehole. The best documented complete sequence is that proved in the Soham Borehole. There, the faunal and lithological sequences can be matched in detail with those proved at outcrop in the East Midlands and in boreholes elsewhere in Fenland. The Lower Oxford Clay contains bituminous mudstone of the type that is worked extensively for brickmaking around Peterborough and Bedford: the Middle and Upper Oxford Clay consist predominantly of calcareous mudstones as at outcrop in western Cambridgeshire.

The sequences at Soham (41.53 m thick), Lakenheath (c. 36 m) and Severals House (c. 39.6 m) are much thinner than that at outcrop in the Huntingdon area, where Woodward (1895, p.56) estimated the thickness of the formation to be about 150 m, but are comparable to those proved in boreholes elsewhere in the eastern part of Fenland (e.g. 40.3 m at Denver Sluice).

DETAILS

Soham Borehole

The most complete Oxford Clay section proved to date in the region is that in the Soham Borehole where the following sequence was proved between 35.36 and 76.89 m:

	Thickness m	Depth m
Upper Oxford Clay		
Quenstedtoceras mariae Zone		
Mudstone, pale and medium grey, calcareous; sparsely shelly but with bivalves locally abundant; rare belemnites, ammonites, gastropods, crustacean debris, and abundant pyritised trails	14.50	49.86
Middle Oxford Clay		
Quenstedtoceras lamberti Zone		
Mudstone, very silty, medium and pale grey, calcareous, cemented to form soft muddy limestone (Lamberti Limestone) in top part; very shelly with common bored and encrusted oysters, and pyritised ammonites; less shelly with depth but with scattered bivalves, ammonites, trails and burrows; oyster-rich bed at base	1.80	51.66
Peltoceras athleta Zone (pars)		
Mudstone, silty, pale grey and brownish grey; calcareous and passing into soft muddy limestones at several levels; common bivalves, rare ammonites, serpulids and pyritised wood	7.32	58.98
Lower Oxford Clay		
Peltoceras athleta Zone (pars)		
Mudstone, pale grey, silty, calcareous interbedded with fissile, brownish grey bituminous mudstone, moderately shelly (calcareous beds) to very shelly (bituminous beds) with common ammonites	11.35	70.33
Erymnoceras coronatum Zone		
Mudstone, weakly calcite-cemented, pale grey, extremely shelly with *Binatisphinctes comptoni* (the Comptoni Bed of the East Midlands) in top-part; resting with bioturbation on thinly interbedded pale grey shelly, calcareous mudstone and brownish grey very shelly, bituminous mudstone with bivalves and ammonites	c.4.8	c.75.1
Kosmoceras jason Zone		
Rhythmic alterations of calcareous mudstone and very shelly bituminous mudstone; common bivalves and ammonites; extremely shelly mudstone at base crowded with oysters, ammonites and belemnites, resting on an irregular, apparently erosive surface	c.1.7	c.76.8
Sigaloceras calloviense Zone (pars)		
Mudstone, bituminous, brownish grey; very shelly with bivalves, ammonites, belemnites and fish debris; bioturbated junction with Kellaways Rock	0.10	76.89
		RWG

The Lower Oxford Clay consists predominantly of fossiliferous brownish grey, fissile mudstones, with subsidiary beds of pale grey, calcareous mudstone in which the fauna is preserved as clay casts, some with original shell material, and some with partial pyritisation. The ammonite fauna, usually crushed flat, is dominated by *Kosmoceras* and a full zonal sequence from the *S. calloviense* Zone to the *P. athleta* Zone is proved. Other very common elements of the macrofauna are the bivalves *Bositra buchii* (Roemer), *Meleagrinella* and nuculoids (particularly *Mesosaccella*), the last named occurring in abundance in shell beds (sometimes pyritised). One of these nuculoid shell beds contains the ammonite *Binatisphinctes comptoni* (Pratt) and is known throughout the East Midlands, where it marks the top of the *E. coronatum* Zone, as the Comptoni Bed. In the borehole, this bed is underlain by pale, very calcareous nuculoid-rich mudstones with thin cementstone bands; a similar sequence is present at this stratigraphical level in brick-pits at Bedford (Callomon 1968, p. 281). *Grammatodon*, *Isocyprina* and *Thracia* are also present, but *Gryphaea* is generally rare. The serpulid *Genicularia vertebralis* (J. de C. Sowerby) is common at some levels, and also the gastropod *Procerithium*. Belemnites (*Cylindroteuthis* and *Hibolites*) also occur.

The Middle Oxford Clay has a distinctive fauna characterised by partially pyritised ammonites including *Quenstedtoceras*, *Kosmoceras*, *Alligaticeras* and *Hecticoceras*, and indicates the presence of the *P. athleta* Zone (below) and the *Q. lamberti* Zone (above) of the Upper Callovian. The boundary between the zones is taken at a *Gryphaea*-rich bed which approximately coincides with the oldest occurrence of *Quenstedtoceras*. The top of the *lamberti* Zone is taken at the top of a fossiliferous siltstone (the Lamberti Limestone) which occurs at this level elsewhere in the Midlands. The non-ammonite macrofauna is transitional between that of the Lower and Upper Oxford Clay, with *Bositra buchii*, *Gryphaea* (often encrusted with foraminifera), nuculoids, *Oxytoma*, rare *Pinna*, *Thracia* and *Dicroloma*. A form that is characteristic of the Lower Oxford Clay, but which occurs in the lower part of the Middle Oxford Clay, is the serpulid *Genicularia vertebralis*. Shell debris/spat and pyritised trails are common throughout.

The Upper Oxford Clay consists of uniform, pale grey calcareous mudstones with a mainly pyritised fauna which includes small ammonites (*Cardioceras*, *Quenstedtoceras*, *Euaspidoceras*, *Hecticoceras* and *Peltoceras*) indicative of the *Q. mariae* Zone of the Lower Oxfordian. Both subzones of the *mariae* Zone (those of *Cardioceras scarburgense* and *C. praecordatum*) are recognised; the boundary between them is taken at about 41.3 m. Bivalves are common at many levels and include arcids, the nuculoids *Mesosaccella* and *Dacryomya*, *Oxytoma*, *Pinna* (common at some levels), *Protocardia*, *Modiolus*, *Thracia*, *Nicaniella* (*Trautscholdia*) and *Gryphaea*, the last named commonly associated with cementstone bands and encrusted with foraminifera. Crustaceans, the gastropod *Dicroloma* and rare small belemnites (*Hibolites*) are also present.

BMC

Methwold Common and Severals House boreholes

Re-examination of the few specimens that have survived from the Methwold Common and Severals House boreholes suggests that they penetrated about 31.1 m (incomplete) and 39.6 m of Oxford Clay respectively. Pringle's logs (1923, pp. 129–131) suggest that the *jason* to *mariae* zones were present at Methwold Common in lithologies similar to those proved at Soham. Both the Comptoni Bed and the Lamberti Limestone appear to have been present. The sequence at Severals House, although complete, is less well documented than that at Methwold Common but is also probably similar to that at Soham.

Lakenheath Borehole

The geophysical logs of the Lakenheath Borehole indicate that about 36 m of Oxford Clay were penetrated, although only the highest 0.4 m was cored; it consists of calcareous mudstone similar to that of the Upper Oxford Clay elsewhere in the region.

WEST WALTON BEDS

The West Walton Beds crop out about 2 km beyond the western boundary and 3 km beyond the southern boundary of the Ely district and were reached in some of the deeper water boreholes in the area. The full thickness of the formation was penetrated in the Methwold Common, Severals House, Lakenheath and Soham boreholes, but only in the last of these was the sampling sufficiently complete for the lithological and faunal sequence to be determined.

Throughout much of Fenland, including a large part of the Ely district, the West Walton Beds consist of siltstones and silty mudstones with muddy limestone concretions at many levels. The formation has been divided into 16 distinctive beds (WWB 1 to WWB 16) on the basis of a combination of lithological and faunal characters (Gallois and Cox, 1977); it overlies the Oxford Clay with minor unconformity. At their maximum development, in the northern part of Fenland, the West Walton Beds are predominantly argillaceous and about 15 m thick. South-eastwards from there, the sequence becomes thinner and more calcareous as the London Platform is approached. The maximum thickness in the Ely district is probably in the March area and is about 14 m. Boreholes in the eastern part of the district have proved argillaceous West Walton Beds from 4 to 8 m, in thickness. At Lakenheath, the mudstones at this stratigraphical level are markedly more calcareous than elsewhere in Fenland. At Soham, they pass laterally into oolitic and pelletal limestones that have been interpreted as near-shore deposits laid down in a lagoon lying between an oolite shoal/patch reef (the Upware Limestone of the Cambridge district) and the London Platform land area (Figure 9).

RWG

DETAILS

Soham Borehole

The following sequence was proved in the Soham Borehole between 24.5 m and 35.36 m:

	Thickness m	Depth m
Cardioceras tenuiserratum Zone		
Siltstone, clayey, pale grey calcareously cemented; passing down into clayey silts with scattered oncoids (algal pellets) that become more common with depth; passing down into	0.89	25.40
Interbedded oncoid-rich (more than 50%), pale grey calcareous mudstone and oncoid-rich, shell fragmental limestone	1.57	26.97
Mudstone, silty and clayey siltstone; calcareously cemented in part; sparsely shelly with *C. tenuiserratum*; bioturbated junction with	1.10	28.07
Cardioceras densiplicatum Zone		
Mudstone, silty, pale grey becoming darker with depth; perisphinctid ammonites common; large, serpulid-encrusted and bored oysters at base rest on an irregular, burrowed surface of Oxford Clay	7.29	35.36

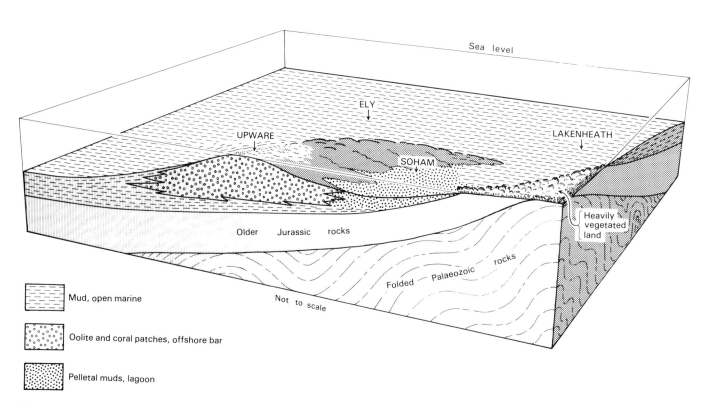

Figure 9   Presumed depositional environments of the West Walton Beds of the district

This sequence is of particular interest because it contains shallow water limestones underlain and overlain by typical argillaceous West Walton Beds lithologies and enables a reconstruction of the environment to be attempted. These limestones probably pass laterally into the oolitic and coralline limestones of the nearby Upware area (the Upware Reef of Arkell, 1933, pp.415–417). The lithological and faunal evidence from the borehole suggests that the lowest part of the West Walton Beds (WWB 1 to WWB 6) is either absent or is combined with Bed WWB 7 to form a condensed basal bed. Representatives of Beds WWB 8 to WWB 11 are present: the last named contains the zonal index *C. tenuiserratum*. Above this, algal pellets are common and indicate deposition within the photic zone, possibly within the intertidal zone. Taken together, the West Walton Beds sequences at Lakenheath and Upware suggest deposition in an open-marine environment close to a shoreline. The Soham sequence probably represents deposition in a partially restricted lagoon protected on its seaward side by an offshore oolite bar (Figure 8).

Methwold Common Borehole

Pringle (1923, p.128) recorded 15.47 m of 'Corallian' in the Methwold Common Borehole between 57.43 and 72.90 m. Specimens in the B.G.S. collection suggest, however, that the Ampthill Clay extends from about 63.1 to 84.7 m, and the West Walton Beds from there to 90.4 m. Very silty, interburrowed, plant-and foraminifera-rich clays, probably West Walton Beds, occur from 85.0 to 88.7 m. The junction of the West Walton Beds and the Ampthill Clay has, therefore, been taken, on the basis of Pringle's log, at the top of a limestone band at 84.7 m. Below this depth, the sequence is problematic. Dark grey clays, unlike any of the West Walton Beds lithologies of the Fenland area, occur at 91.4 to 91.6 m. The base of the West Walton Beds has been taken at 90.4 m on the basis of Pringle's (1923, p.129) lithological description, although the classification is clearly still open to doubt.

Severals House Borehole

No specimen of West Walton Beds has survived from the Severals House Borehole. The beds between 97.5 m and 105.2 m have been assigned to the West Walton Beds on the basis of Pringle's borehole log which records alternations of clay and limestone for this interval.

Lakenheath Borehole

In the Lakenheath Borehole, an attenuated West Walton Beds sequence is present between 91.8 m and 95.9 m, and consists of pale grey, very calcareous soft mudstones, silty mudstones, and dark grey, plant-rich, very silty mudstones with calcareously cemented bands (muddy limestones) at several levels. The sparse fauna includes *Gryphaea* encrusted by *Bullapora* and other foraminifera, *Chlamys*, cardioceratid and perisphinctid ammonite fragments, *Dicroloma* and plant debris. The presence of relatively common *Cardioceras tenuiserratum* (Oppel) in the top part of the sequence indicates the presence of the Middle Oxfordian *Cardioceras tenuiserratum* Zone. The basal bed of the West Walton Beds in the borehole contains burrowfill concentrations of foraminifera and plant debris that penetrate the top of the Oxford Clay, a feature which is common at this stratigraphical level elsewhere in Fenland.   BMC,RWG

AMPTHILL CLAY

The Ampthill Clay crops out over about one third of the Ely district but is poorly exposed: beneath a large part of the area it is overlain by the Quaternary deposits of southern Fenland. The formation comes to the surface in the low Fenland 'islands' on which Chatteris, March, Wimblington, Manea and Pymore are built, and on the western flanks of the much larger 'Isle of Ely' between Haddenham, Sutton, Mepal and Coveney.

There is no natural or permanent exposure of Ampthill Clay within the district and, because of this, there has been no comprehensive account of its local stratigraphy. A number of workers have described individual sections or have made faunal collections from them: Dr G.A. Kellaway and Mr S.C.A. Holmes collected material from temporary sections in the Chatteris, Wimblington and Manea areas, and the fauna from some of these was commented on by the late Dr Arkell. Hancock (1954) described an Ampthill Clay fauna from exposures in the Cam valley (Cambridge District), and Forbes (1960) described sections in the formation at Mepal and Sutton. Pringle (1923) described the Ampthill Clay succession in the Methwold Common Borehole. In recent years, large parts of the Ampthill Clay have been continuously cored in boreholes adjacent to the district and its stratigraphy is now known in detail.

The Ampthill Clay consists largely of shelly, soft, dark grey mudstones, slightly silty mudstones and pale grey, calcareous mudstones. Doggers and thin beds of muddy limestone (cementstone) occur at several levels, usually in association with the more calcareous mudstones, and thin beds of clay ironstone and organic (kerogen-rich) mudstone are also present. Pyrite occurs throughout the formation and bands of phosphatic nodules are common at several levels. At outcrop, the Ampthill Clay weathers to a sticky grey to yellowish brown clay containing common crystals of selenite ($CaSO_4.2H_2O$) that have formed by the interaction of shelly fossils and the oxidation products of pyrite.

In the present district, the base of the Ampthill Clay has been taken at an upward lithological change from the medium and pale grey silty calcareous mudstones of the West Walton Beds into medium and dark grey slightly silty mudstones; there is an associated change in the fauna from rare small ammonites preserved as clay casts to abundant, more varied ammonites with white calcite shells preserved. These changes are persistent throughout the district and over a large part of eastern England from Cambridgeshire to Lincolnshire.

The formation can be divided into three on the basis of gross lithology. The lowest part is generally slightly silty and forms a passage down into the West Walton Beds; the middle part consists mostly of smooth textured clays; and the highest part is characterised by calcareous mudstones with several erosion surfaces marked by phosphatic pebbles and oysters, and by silty lithologies similar to those of the West Walton Beds. In the Cambridge district, the pebble beds in the upper part of the Ampthill Clay were mapped out as 'phosphate beds' (Holmes and others, 1965) although the total amount of phosphate in them is very small.

The Ampthill Clay thins southwards at outcrop across the Ely district from an estimated 50 m in the Welney area to 33 m at Haddenham. This thinning is due almost entirely to attenuation in the upper part of the formation where a number of erosion surfaces combine: it probably reflects a change to more turbulent environments as the London Platform is approached. The Ampthill Clay has an extensive subcrop in the district and is probably present everywhere beneath the Kimmeridge Clay, except for the small area south-east of a line from Lakenheath to Soham, where it has been removed by erosion in the early Cretaceous. Within the subcrop, it thins eastwards to about 20 m in the Methwold Fens area, again reflecting the influence of the London Platform.

The Ampthill Clay is composed of clay minerals, quartz and calcium carbonate with lesser amounts of pyrite, phosphate, siderite and organic matter. The clay-mineral content ranges from about 25 to 60 per cent and consists of clay mica, probably largely illite, kaolinite and chlorite (Merriman and Strong in Gallois, 1979a). Small amounts of vermiculite have been recorded (Perrin, 1957) from surface exposures but this may be a weathering product. The silt content of the Ampthill Clay consists mostly of quartz: the total amount of crystalline quartz, as detected by X-ray diffraction analysis, ranges from about 7 to 22 per cent. The calcium carbonate content varies from less than 15 per cent in the dark grey mudstones to 30 to 50 per cent in the pale grey mudstones. Cementstone doggers with calcium carbonate contents of up to 90 per cent commonly occur in the pale grey mudstones and the silty mudstones: they appear to have formed as early diagenetic concretions by the migration of calcium carbonate from the shelly fauna and, in the pale grey mudstones, from a fine calcite mud which is probably largely biogenic and derived from the disintegration of coccoliths, calcispheres, foraminifera and bivalves.

The Ampthill Clay is characterised by marine faunas made up largely of ammonites, bivalves and foraminifera; these suggest deposition in a shallow shelf-sea. Oysters, gastropods, echinoderm, crustacean and plant fragments are common at certain levels, suggesting a near-shore environment (Plate 2). The phosphatic pebble beds in the upper part of the Ampthill Clay probably indicate current-agitated conditions produced by a shallowing of the sea as a result of reactivation of the nearby land area. In contrast, smooth-textured clays with crushed, thin-shelled ammonites and bivalves are also present, and these probably indicate placid, more offshore or deeper water conditions. The relationship of the more common faunal and lithological associations to their presumed environments of deposition is summarised in Figure 10.

The formation has been zoned on the basis of species of the ammonites *Amoeboceras* and *Cardioceras* (Sykes and Callomon, 1979), and has been divided in the Fenland area into 42 beds on the basis of combined faunal and lithological characters (Gallois and Cox, 1977). The generalised vertical section for the Ampthill Clay of Fenland showing the lithologies, main faunal features, zones/subzones, and bed numbers is shown in Figure 11 (based on Gallois and Cox, 1977, fig.2).

Most of the fauna and many of the lithological characters are destroyed at outcrop by weathering. The pyritic fossils are usually lost, and the only calcitic forms that are commonly preserved are the thicker-shelled oysters. A more complete fauna is preserved in the cementstones, but this is dominated by bivalves and facies-sensitive perisphinctid ammonites, and can give a biased impression of the fauna of the formation. The zonal ammonites are common in unweathered material but are rarely preserved at outcrop. Material collected from cored boreholes in and adjacent to the present district, together with museum collections of material from former brick pits, borrow pits, gravel pits and temporary sections in the deeper drains, have enabled the detailed stratigraphy to be determined.

The Ampthill Clay crops out on the lower slopes of the Fenland islands at Wimblington, Honey Hill, Langwood Hill, and Manea and occurs at shallow depth beneath Quaternary deposits at Block Fen, Byall Fen and Langwood Fen. It crops out as a continuous belt on the slopes along the western edge of the 'Isle of Ely' between Haddenham and Coveney and forms small outliers on the 'islands' of Pymore and Butcher's Hill. Northwards from Butcher's Hill, the subcrop of the formation beneath Quaternary deposits has been traced by boreholes as a broad belt running beneath the Norfolk and Lincolnshire parts of Fenland.

In the present district, the Ampthill Clay is overlain with minor unconformity by the Kimmeridge Clay. Evidence from adjacent areas suggests that the Kimmeridge Clay oversteps the Ampthill Clay in an easterly direction but that the rate of overstep is slow (less than 1 m per 10 km). In the adjacent Cambridge district, Holmes (in Worssam and Taylor, 1969, p.72) recorded folds in the Ampthill and Kimmeridge clays that are planed off by the basal unconformity of the Lower Greensand. Synclinal pockets of Ampthill Clay may, therefore, be preserved beneath the Lower Greensand in the south-eastern part of the present district, but none has yet been proved in boreholes.   RWG

DETAILS

In the following descriptions of the sections the bed numbers (AmC = Ampthill Clay) refer to Figure 11. Generalised descriptions of the individual beds are given elsewhere (Cox and Gallois in Gallois, 1979a). The outcrop and subcrop of the Ampthill Clay and the sites of exposures and boreholes referred to are shown in Figure 12.

Chatteris – Wimblington

The highest beds of the West Walton Beds and the lowest beds of the Ampthill Clay (Beds AmC 1 to 4) were proved in the spoil dredged from the 'Gault' drain at Chatteris [385 856 to 386 863], about 1.5 km west of the present district. It seems likely, therefore, that the oldest Ampthill Clay exposed in the Ely district is that in ditches [408 865] close to the western boundary of the district near Delve Farm and in the floor of Vermuyden's Drain near Robinson's Farm [407 877], Chatteris. The faunal assemblage from the lower part of the Ampthill Clay in the 'Gault' drain includes the ammonites *Cardioceras* cf. *cawtonense* (Blake & Hudleston), *C. kokeni* Boden, *C. maltonense* (Young & Bird), *C. tenuiserratum* (Oppel) with *Grammatodon*, *Oxytoma*, rare serpulids and a belemnite fragment (*Pachyteuthis?*).

The most distinctive beds of the Ampthill Clay of the district are Bed AmC 12, a pale grey calcareous mudstone with cementstone nodules (locally passing into a tabular cementstone) and a rich perisphinctid fauna, and Bed AmC 15, a dark grey mudstone with a

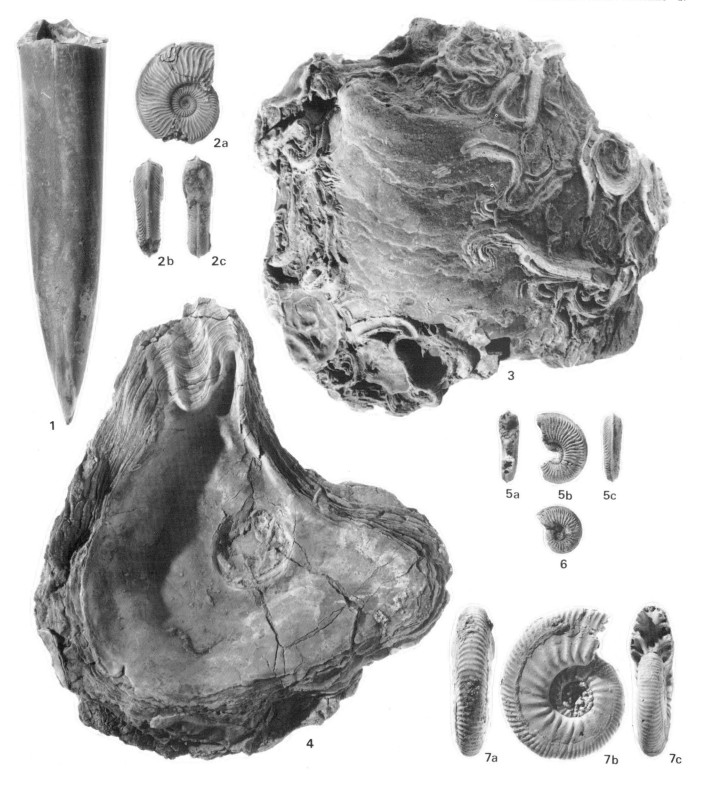

Plate 2   Fossils from the Ampthill Clay (all × 1)

1 *Pachyteuthis abbreviatus* (Miller); Sixteen Foot Drain, Stonea (SM J.24595)

2a–c *Amoeboceras newbridgense* Sykes & Callomon; Bed AmC 15, *glosense* Zone; Old Fordey Farm, Stretham (BGS FD2094)

3 *Gryphaea dilatata* J. Sowerby, encrusted with serpulids; Bed AmC 13, *tenuiserratum* Zone; Block Fen gravel pit, Mepal (BGS RWG63)

4 *Deltoideum delta* (Wm Smith); Beds AmC 25–27, *serratum-regulare* zones; Bury Lane, Sutton (SM X.7810)

5a–c, 6 *Amoeboceras* cf. *ilovaiskii* (Sokolov); Bed AmC 15, *glosense* Zone; Block Fen gravel pit, Mepal (BGS RWG78 and RWG72)

7a–c *Decipia sp.*, pyritised inner whorls; Bed AmC 15, *glosense* Zone; Old Fordey Farm, Stretham (SM J.35807)

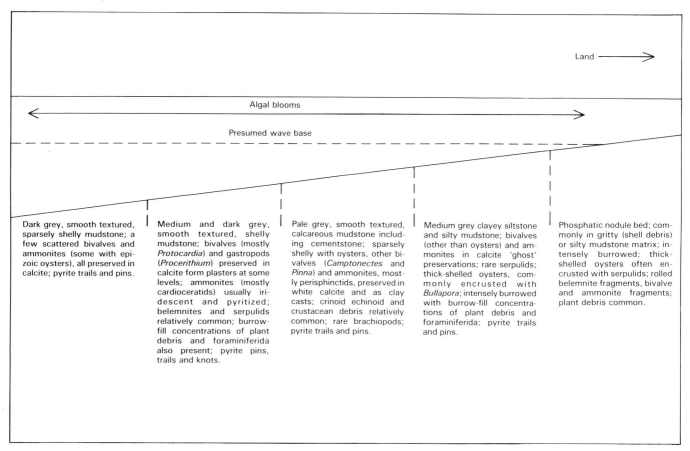

Figure 10 Relationship of Ampthill Clay lithologies to their presumed depositional environments

rich but stunted *Amoeboceras* fauna. A section in the bank of the Sixteen Foot Drain at Honey Bridge, Wimblington, [435 895] showed 1.8 m of grey clay with *Gryphaea dilatata* J. Sowerby heavily encrusted with serpulids overlying 0.9 m of pale grey calcareous clay and cementstone with *Chlamys*, *Entolium*, *Nanogyra* and assorted perisphinctid fragments including nuclei, inner whorls and body chambers. The larger ammonite fragments were identified by Arkell (1944, MS) as *Decipia* and/or *Pomerania*. The faunal assemblage shows the pale clay to be Bed AmC 12; the presence of *Gryphaea* in the overlying clay suggests that the basal part of Bed AmC 13 is also present. Similar lithologies and faunas were collected from the spoil from Vermuyden's Drain at Horseway, Wimblington [429 871 and 427 871]. There, pale calcareous clay with *Decipia* and/or *Pomerania* (loc. cit.) and *Entolium* (Bed AmC 12) were overlain by grey clay with serpulid- encrusted *Gryphaea dilatata* (Bed AmC 13).

Mepal

A section in an old brick pit at Fortrey's Hall, Mepal [444 827], showed 0.4 m of pale calcareous clay, similar in lithology and fauna to Bed AmC 12 at Honey Bridge and Horseway, with *Astarte*, *Entolium*, *Pinna*, *Thracia?*, *Dicroloma* and fragments of perisphinctid ammonites. This clay was overlain by up to 2.7 m of grey clay with *G. dilatata*, *Nanogyra* and small pyritic ammonites, including perisphinctid nuclei in the lower part and *Amoeboceras* in the upper part. This assemblage suggests the presence of Bed AmC 13 and the basal part of Bed AmC 15.

An ammonite fauna characteristic of Bed AmC 15, consisting of *Amoeboceras ilovaiskii* (Sokolov), *A.* cf. *transitorium* Spath and indeterminate perisphinctid nuclei, all in pyritic preservation, was col-

lected during the present study from Ampthill Clay spoil in the floor of a working gravel pit at Block Fen, Mepal [430 836]. The spoil also included serpulid-encrusted *G. dilatata* (Bed AmC 13) and slabs of siderite mudstone up to 0.1 m thick with an irregular (presumed lower) surface and a flat, serpulid- and oyster-encrusted upper surface with lithophagid borings (Bed AmC 16).

The Great Ouse River Authority borrow-pit at Toll Farm, Mepal [436 814], formerly exposed about 5.5 m of clay with *Amoeboceras*, rare *Decipia*, common perisphinctid nuclei, belemnites and numerous bivalves (Forbes, 1960, p. 238). Specimens preserved in the B.G.S. and Sedgwick Museum collections include *G. dilatata* encrusted with *Placunopsis* and serpulids (Bed AmC 13), body chambers of *Decipia* infilled with very pale, partially phosphatised mudstone (Bed AmC 14), clays with abundant *Amoeboceras* including *A. ilovaiskii* (Bed AmC 15) and common *Deltoideum delta* Wm Smith (probably Bed AmC 20). A number of densely phosphatised *Amoeboceras* fragments collected loose from the drift at the top of the pit were probably derived from Bed AmC 30 or a younger bed of the Ampthill Clay.

Upware

Worssam and Taylor (1969, p.19) recorded a stratigraphically important section in the inlier of Ampthill Clay brought up by the Upware anticline (Cambridge district). They described 7.2 m of dark grey clay with cementstone and ironstone nodules in a borrow pit at Old Fordey Farm, West Hill [542 749], close to the southern boundary of the present district. An extensive fauna was collected, and lithological and faunal representatives of Beds AmC 12 to 16 are present. Pale yellowish grey fossiliferous cementstones, from which Hancock (1954, p. 251) recorded *Decipia* and other perisphinctids

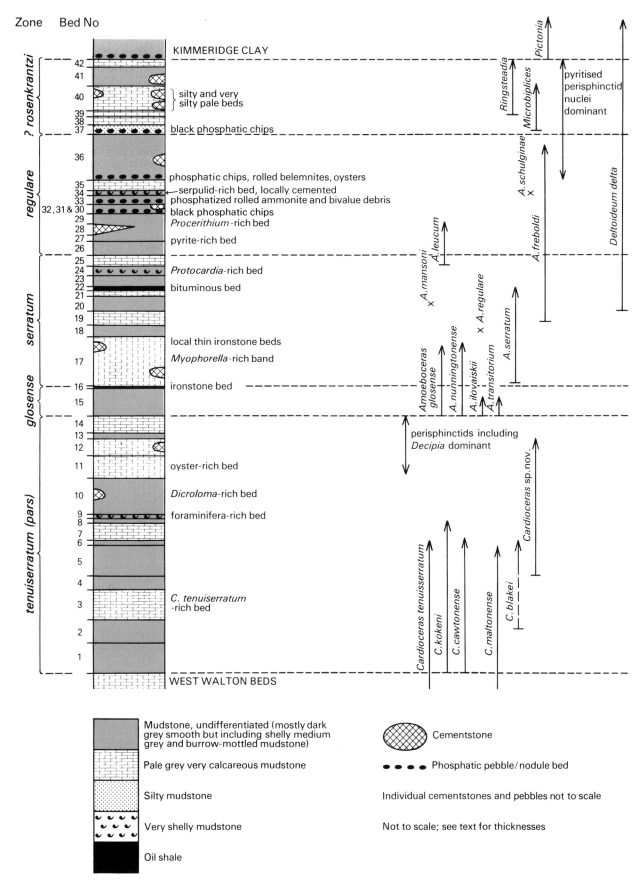

Figure 11 Generalised vertical section of the Ampthill Clay of the district

(Bed AmC 12), were overlain by silty clays with *Decipia*, *Chlamys*, small gastropods (*Procerithium?*) and common *G. dilatata* heavily encrusted with serpulids (probably Bed AmC 13), and clays with pyritised *Amoeboceras* including *A. glosense* (Bigot & Brasil) and *A. ilovaiskii* (Sokolov) (Bed AmC 15). The clay ironstone nodules may indicate the presence of Bed AmC 16.

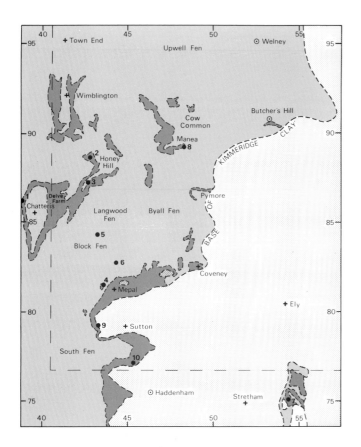

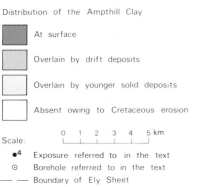

Distribution of the Ampthill Clay

At surface

Overlain by drift deposits

Overlain by younger solid deposits

Absent owing to Cretaceous erosion

Scale: 0   1   2   3   4   5 km

●4   Exposure referred to in the text

⊙   Borehole referred to in the text

— —   Boundary of Ely Sheet

Figure 12   Ampthill Clay localities referred to in the text

1	'Gault' drain, Chatteris
2	Honey Bridge, Wimblington
3	Horseway, Wimblington
4	Toll Farm, Mepal
5	Block Fen, Mepal
6	Fortrey's Hall, Mepal
7	Old Fordey Farm, West Hill
8	brick pit, Manea
9	Bury Lane, Sutton
10	Salmon's Farm, Haddenham

Manea

A brick pit at Manea [483 892] formerly exposed up to 4.9 m of dark grey calcareous clay with a band of cementstone nodules some 0.9 to 1.8 m from the top of the section; the section includes representatives of Beds AmC 17 to AmC 19. The faunal assemblage of Beds AmC 17 and 18 includes perisphinctid and *Amoeboceras* fragments (including *A*. cf. *regulare* Spath, *A*. cf. *glosense*, *A. nunningtonense* Wright and *A*. cf. *serratum* (J. Sowerby)), the bivalves *Camptonectes*, *Grammatodon*, *Liostrea* and *Thracia*, and *Procerithium*. A collection of fish fragments, identified by the late Dr E.I. White as the bony fish *Lepidotus* cf. *leedsi* A.S. Woodward, came from the lower part of the section (Bed AmC 17). A shell plaster occurs at about 2.4 m from the top with *Dicroloma*, *Camptonectes*, *Oxytoma* and echinoid plates, and marks the base of Bed AmC 18. The fauna of Bed AmC 19, which occupies the highest 1.8 m of the section, includes *Amoeboceras* (including *A*. cf. *serratum*) and perisphinctid fragments, an echinoid spine, tiny *Procerithium* (common), the bivalves *Camptonectes*, *Grammatodon*, *Thracia* and small oysters, and a fragment of belemnite (*Cylindroteuthis?*).

Haddenham – Sutton

Spoil from ditches in the Ampthill Clay near Salmon's Farm, Haddenham [447 769, 448 769 and 457 777], yielded encrusted and bored specimens of *Gryphaea dilatata*. The junction of the Ampthill and Kimmeridge clays was formerly exposed in a large borrow pit (Figure 13), at Bury Lane, Sutton [432 793], made by the Great Ouse River Authority to provide embanking materials (Forbes, 1960, p.234). The following section was recorded by Dr Forbes and the late Dr Arkell. The faunal and bed classifications are based on the present authors' re-examination of collections from this locality held at the B.G.S. and the Sedgwick Museum.

Kimmeridge Clay		see p.36 for details
Bed No.		*Thickness* m
AmC 40:	Mudstone and silty mudstone, mottled grey and yellow; seepage at base	0.8
AmC 38 & 39:	Mudstone, dark grey; selenitic where weathered, shelly with crushed *Thracia* and rare pyritised perisphinctid inner whorls; large plesiosaur bone	1.8
AmC 30 to 37: condensed	Siltstone crowded with small black phosphatic pebbles and larger (up to 6 cm across) bored and serpulid-encrusted, cream-coloured phosphatic pebbles, some enclosing bivalves, wood and fragments of *Amoeboceras* and *Microbiplices?*; rolled and bored belemnites common; matrix fossiliferous with small belemnites, *Deltoideum delta*, echinoid spines and wood fragments	0.08
AmC 29:	Mudstone, shelly with crushed bivalves, including *Protocardia* and *Thracia*, preserved in white calcite and with iridescent ammonites including *Amoeboceras*	1.1
AmC 28:	Oyster lumachelle composed of *Deltoideum delta*; lenticular bed locally swelling to form a bioherm 1 m thick and 6 m across in one small area; mostly up to	0.3
AmC 25 to 27:	Mudstone, dark grey and dark, slightly greenish grey; shelly with crushed bivalves preserved in white calcite including *Deltoideum delta* and *Thracia*	

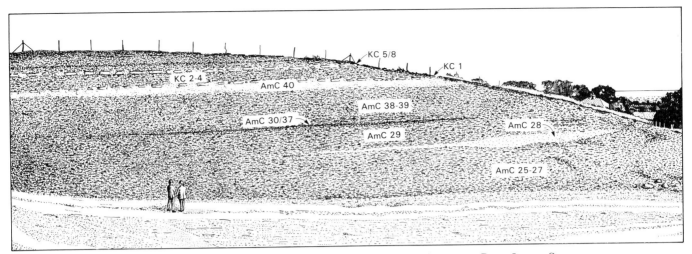

Figure 13 Ampthill Clay – Kimmeridge Clay junction and topographical feature at Bury Lane, Sutton

Thickness
m

depressa; common well preserved belemnites including *Pachyteuthis* and *Belemnop-
sis*; abundant crushed iridescent ammonites, some with phosphatised or
pyritised body chambers, mostly
Amoeboceras (including *A.* cf. *leucum* Spath,
A. cf. *mansoni* Pringle and *A. serratum*)
and also *Perisphinctes*   3.8

A thick phosphatic nodule bed, probably that classified in the above
section as Beds AmC 30/37, was also recorded in a ditch near Haddenham End Field [459 769] (Worssam and Taylor, 1969, p.22); it
contained numerous fragments of rolled and bored bivalves and
Amoeboceras preserved in shiny black phosphate. These *Amoeboceras*
indicate derivation from the *Amoeboceras regulare* Zone and/or
Amoeboceras rosenkrantzi Zone: one specimen shows striking rur-
siradiate ribbing, a feature which is characteristic of the younger
zone. The remaining derived fauna includes *?Isocyprina
(Venericyprina)*, rolled belemnites including *Cylindroteuthis?*, and a
perisphinctid ammonite fragment preserved in cream-coloured
phosphate.

Boreholes

The Ampthill Clay has been proved beneath Quaternary deposits in
several shallow boreholes in the district but has been either too
deeply weathered or insufficiently sampled for the horizons to be
subsequently determined. Boreholes at White Gate Farm, Stonea
[4506 9446]; Copes Hill Farm, Welney [5265 9478]; Wenny Farm,
Chatteris [4135 8487] and Oxlode Pumping Station [4824 8579]
penetrated up to 5 m of Ampthill Clay.'

The full thickness of the formation was penetrated in the
Methwold Common Borehole (Pringle, 1923). The borehole was
continuously cored, but few specimens have survived and it is now
impossible to subdivide the formation in detail. Pringle (1923, p.
128) recorded 15.47 m of 'Corallian' clay but re-examination of the
specimens in the BGS collection suggests that the Ampthill Clay extends from about 63.1 m to 84.7 m, a total thickness of 21.6 m. A
specimen from 63.1 m shows medium grey silty mudstone interburrowed with smooth-textured pale grey mudstone and is probably
from close to the Ampthill Clay – Kimmeridge Clay junction.
Typical Ampthill Clay lithologies occur at 64.6 m and 71.3 m. The

junction of the West Walton Beds and the Ampthill Clay has been
taken at the top of a limestone band at 84.7 m on the basis of Pringle's log.

Other boreholes close to the district boundary have proved all or
part of the Ampthill Clay. Cored boreholes at Denver Sluice [TF
5911 0106] and Severals House, Methwold [6921 9639], both on
Sheet 159, proved complete Ampthill Clay sequences 35.71 m and
20.4 m thick respectively (Gallois and Cox, 1977); the difference in
thickness is due to easterly attenuation within the formation. Site-
investigation boreholes at March [416 968], also on Sheet 159, proved up to 7 m of fossiliferous grey mudstones and calcareous
mudstones with representatives of Beds AmC 5 to 11 (*Cardioceras
tenuiserratum* Zone, ? *C. blakei* Subzone). The fauna is dominated by
bivalves and cardioceratid ammonites including *C. cawtonense* (Blake
& Hudleston), *C. kokeni* (Boden), *C. maltonense* (Young & Bird) and
C. tenuiserratum (Oppel). *Perisphinctes* and perisphinctid fragments,
Gryphaea dilatata, *Oxytoma*, *Dicroloma*, *Procerithium* and clusters of serpulids are also present.

The full thickness of the Ampthill Clay was cored in the BGS
borehole at Haddenham [4662 7554] on the northern edge of Sheet
188. There, about 33 m of mudstone and calcareous mudstone was
proved between the West Walton Beds (formerly Elsworth Rock
Group) and the Kimmeridge Clay (Horton *in* Institute of
Geological Sciences, 1971, p.104; Medd, 1979). The lower part of
the sequence (Beds AmC 1 to 29) is similar to that proved elsewhere
in Fenland. The higher part (Beds AmC 30 to 40) resembles the
condensed sequence proved at this level at Bury Lane, Sutton. Two
phosphatic pebble beds were proved in the upper part of the Ampthill Clay in the borehole; a thick, composite pebble bed at 36.07 to
36.42 m probably represents Beds AmC 30 to 36, and a pebble bed
at 35.07 m is probably that at the base of Bed AmC 37. The Kimmeridge Clay rests at 33.40 m on an irregular, intensely burrowed
surface of pale grey mudstone (Bed AmC 40).

The basal beds of the Ampthill Clay, overlain unconformably by
Lower Greensand, were cored between 88.4 m and 91.8 m in the
Lakenheath Borehole. Pale grey smooth-textured calcareous
mudstones with *Cardioceras tenuiserratum* (Beds AmC 1 to 4) rest at
about 91.8 m on slightly silty and silty mudstones with several
cementstone horizons (West Walton Beds).

Ampthill Clay is absent, through erosion at the base of the Lower
Greensand, beneath much of the eastern part of Sheet 188 and the
south-eastern part of the Ely district. The BGS cored boreholes at
Cambridge and Soham (Holmes *in* Worssam and Taylor, 1969,
p.7) and Ely-Ouse Borehole 12 [6962 7858] near Lakenheath, proved Lower Greensand resting on West Walton Beds.   BMC, RWG

KIMMERIDGE CLAY

In the southern part of the district, between Haddenham End and Stuntney, the Kimmeridge Clay has a wide, largely drift-free outcrop. Northwards from there, the outcrop becomes progressively more drift-covered until, in the area north of Littleport, Kimmeridge Clay comes to the surface only in the 'island' of Southery. Much of the 'island' of Ely and those of Littleport, Shippea Hill and Southery are underlain by Kimmeridge Clay. The formation has been proved in boreholes at shallow depths beneath Recent deposits under Middle Fen, Padnal Fen, Burnt Fen, Fodder Fen, Southery Fens and Hilgay Fen. Between there and a line running from Soham to Lakenheath, it has been proved at greater depths, beneath either Sandringham Sands or Carstone. To the south-east of this line, the Kimmeridge Clay has been removed by erosion during the late Jurassic or early Cretaceous.

Despite the present paucity of exposure within the district, most of the sequence has been available at some time for collecting in brick or borrow-pits. The more calcareous parts of the Kimmeridge Clay were formerly worked for brick-making at Haddenham End, Downham, Littleport and Southery. Kimmeridge Clay has also been worked for embanking materials to repair the Fenland drainage system in large borrow-pits at Sutton and Stretham and continues to be worked for this purpose at Ely. Most of these sections are now degraded but faunal and lithological information obtained from them, and from continuously cored boreholes in and adjacent to the district, has enabled the detailed stratigraphy of the formation to be determined.

The Kimmeridge Clay is made up of soft mudstones, calcareous mudstones and kerogen-rich mudstones (oil shales) with lesser amounts of silty mudstone, siltstone and muddy limestone. Pyrite is present throughout and phosphatic pebble beds occur at a few levels near the base of the formation. The mudstones are shelly at most levels and contain a rich marine fauna dominated by bivalves and ammonites, but with gastropods, serpulids, brachiopods, echinoderms and vertebrates common at some levels (Plate 3). As with the Ampthill Clay, much of the faunal and lithological detail of the Kimmeridge Clay is destroyed at outcrop by weathering, and both formations give rise to heavy grey to yellowish brown subsoils rich in selenite.

In the present account, the base of the Kimmeridge Clay has been taken at a lithological and faunal change that occurs at a minor, but widespread, unconformity. At outcrop, between Coveney and Haddenham, the basal beds of the Kimmeridge Clay are condensed compared to other areas of Fenland. Between Coveney and Sutton, the basal bed consists of 0.5 m of very pale grey, silty mudstone containing limestone concretions with common ammonite body chambers, and abundant terebratulid and rare rhynchonellid brachiopods. This distinctive lithology is overlain by about 2 m of mudstone which in turn is overlain by a 0.2 m thick siltstone with phosphatic pebbles and partially phosphatised cobbles of limestone derived from the underlying Kimmeridge Clay or the Ampthill Clay. In the Haddenham area, this siltstone rests locally on the Ampthill Clay. The age of the lowest Kimmeridge Clay, and that of the Ampthill Clay on which it rests, thus varies slightly even within the

small area of outcrop present in the Ely district. At Haddenham, the basal siltstone rests on pale grey mudstones in the upper part of Bed AmC 40: at Sutton, the basal limestone rests on silty mudstones within the lower part of this bed. In both cases, the lithological change is accompanied by an upward change in the ammonite assemblages from one dominated by *Amoeboceras* and *Ringsteadia* to one in which *Pictonia* and *Rasenia* predominate.

Between Sutton and Mepal, the limestone and siltstone bands in the basal part of the Kimmeridge Clay give rise to a well marked feature (Figure 13) and, locally, to a spring line. The highest part of the Ampthill Clay weathers to produce an acid soil, and its junction with the Kimmeridge Clay has been mapped in this area by means of a combination of features including soil fragments, topographic expression and the boundary of the acid soils (Seale, 1975).

In the area between Southery and Methwold Common, that in which the most complete Kimmeridge Clay sequence is present in the district, the formation is about 44 to 46 m thick. In the northern part of the district (north of the Little Ouse River), the formation is unconformably overstepped southwards by the Sandringham Sands in an almost planar manner. South of the Little Ouse, the Lower Greensand rests on the Kimmeridge Clay in a much more irregular fashion, and the formation is completely removed by this erosion south-east of a line from Brandon to Soham. The outcrop and subcrop of the formation, and the age at particular localities of the youngest bed preserved beneath the Sandringham Sands or the Lower Greensand, are shown in Figure 14.

The Kimmeridge Clay of the Ely district is thin in comparison with that of adjacent parts of Fenland. This is due to two factors: attenuation within the formation as it ap-

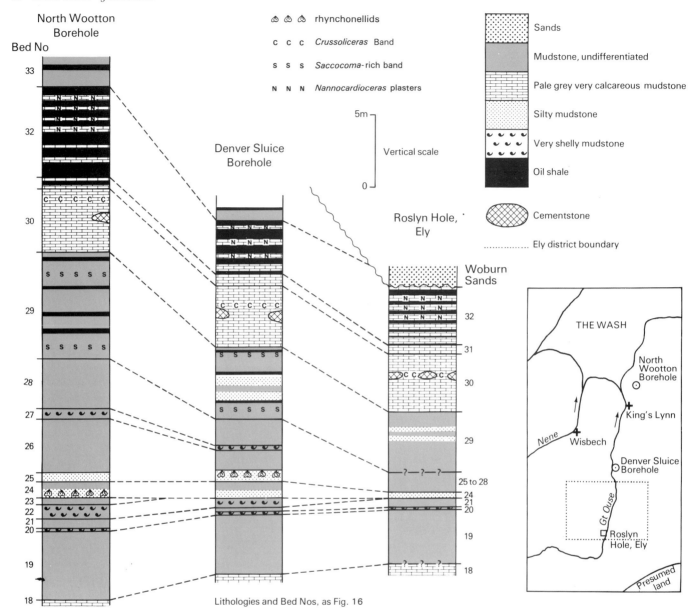

Figure 14 Attenuation in part of the Lower Kimmeridge Clay between The Wash and Ely

proaches the stable area of the London Platform, and pre-Aptian erosion. Although the lithological and faunal sequences within the Kimmeridge Clay of the district can be matched in detail with those of more northerly parts of Fenland, small differences are present at Ely that reflect the proximity of land. These include the presence of minor erosion surfaces and additional thin siltstone horizons (Figure 15). The common occurrence of relatively large pieces of wood in the Kimmeridge Clay at Ely, some bored by bivalves, may indicate a nearby vegetated area.

The Kimmeridge Clay is made up of a sequence of small-scale rhythms, mostly 0.5 to 2 m thick. In the lower part of the formation, these rhythms consist of thin siltstones or silty mudstones overlain by dark grey mudstones and pale grey calcareous mudstones. In the middle and upper parts of the Kimmeridge Clay, they consist of oil shales overlain by dark

grey mudstones and pale grey calcareous mudstones. These rhythms are laterally persistent throughout the district and many can be correlated over distances of tens of kilometres in the adjacent areas. Superimposed on this rhythmic sequence are broader lithological changes, from more to less calcareous, and from more to less oil-shale-rich, which can be regarded as larger scale rhythms and can be correlated throughout the English Kimmeridge Clay.

Using the parameters of colour, texture, grain size and shell content, a number of lithologies can be readily recognised in borehole cores and outcrop sections. The more distinctive lithologies include barren mudstones (pale, medium or dark grey), shelly mudstones (pale, medium or dark grey), silty mudstones, thin limestones (muddy, septarian or coccolith-rich), oil shales (shelly or barren), and thinly interbedded, interlaminated and interburrowed combinations of these.

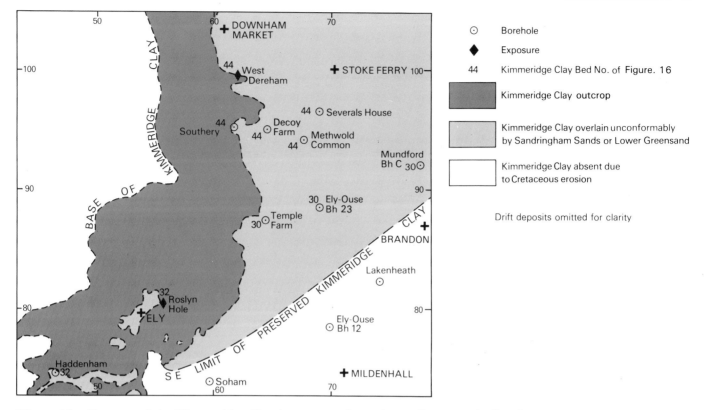

Figure 15   Overstep of the Kimmeridge Clay by younger formations adjacent to the London
Platform

These lithologies are made up of a small number of clastic, biogenic and chemogenic components, which occur in varying amounts. The main clastic components are clay minerals (mostly illite and kaolinite but with small amounts of chlorite), quartz, and reworked biogenic and chemogenic materials in the form of transported shell and plant debris, and phosphatic pebbles. The biogenic components include the calcareous fauna and flora (ammonites, bivalves, brachiopods, crinoids, serpulids, foraminifera, ostracods and coccoliths), the phosphatic fauna (vertebrate debris), and the organic flora (plant debris, spores, pollen and dinoflagellate cysts). The chemogenic components consists of diagenetically formed calcite, kerogen, phosphate, and pyrite.

The gross mineralogy of the commoner Kimmeridge Clay lithologies can be expressed in terms of the four components clay minerals, quartz, calcium carbonate, and kerogen. The typical composition of the most common lithologies, have been determined by X-ray diffraction analysis and can be summarised as follows:

1   dark grey mudstone — clay minerals 45 to 65%; quartz 10 to 30%; calcium carbonate 5% to 20% depending upon shell content; kerogen less than 1%

2   pale grey mudstones — clay minerals 25 to 45%; quartz 8 to 15%; calcium carbonate 25 to 55%; kerogen less than 1%

3   cementstone — clay minerals 10 to 20%; quartz 2 to 6%; calcium carbonate 60 to 90%; kerogen less than 1%

4   oil shale — clay minerals 20 to 40%; quartz 10 to 15%; calcium carbonate 10 to 25%; kerogen 10 to 45%

The Kimmeridge Clay is wholly marine throughout Britain, and at most levels is rich in ammonites and bivalves. The formation has been zoned on the basis of the former (Ziegler, 1962; Cope, 1967, 1978), and has been further subdivided on the basis of a combination of lithological and faunal characters (Gallois and Cox, 1976; Cox and Gallois 1981). This latter classification has been used in the present work to determine the stratigraphical positions of the various outcrop and borehole sections in the Ely district. The lithological sequence, main faunal features, zones and bed numbers are summarized in Figure 16.

The most useful single feature for determining the stratigraphical position of any outcrop or borehole section within the Kimmeridge Clay is the ammonite assemblage. Most sections are sufficiently fossiliferous for the zonal position to be determined; in many cases in the Lower Kimmeridge Clay, the position within the zone (either low or high) can be determined from the ammonites alone. In addition to the zonal index genera, species of *Amoeboceras* (including *Nannocardioceras*), *Aspidoceras* and its aptychal plate *Laevaptychus*, *Crussoliceras*, *Sutneria* and *Xenostephanus* provide useful markers, and enable sub-divisions of zones to be identified with confidence. In boreholes, the zonal boundaries and the positions of selected marker beds can be rapidly identified using a combination of a small number of core specimens and geophysical logs.

Thin marker bands, containing flood occurrences of coccoliths, crinoids, certain species of bivalve, brachiopods or serpulids, occur at many levels. When combined with the rhythmic variation in lithology that occurs throughout the Kimmeridge Clay, these faunal markers provide a ready means of subdividing the formation. Many of the faunal changes coincide with lithological changes that can be recognised throughout southern England; this suggests that they reflect widespread events affecting the whole of the Kimmeridge Clay basin of deposition.

Sections in most of the Victorian brick pits in the district were described by Roberts (1892) in his classic work on the Upper Jurassic rocks of Cambridgeshire. Re-examination by the present authors of material collected by him and others has enabled these sections to be placed within the modern stratigraphical classification. Rose (1835) noted that 'inflammable schist' [oil shale] occurred in a brick-pit at Southery. Forbes (1960) described the basal beds of the Kimmeridge Clay in a large borrow pit at Sutton, and Worssam and Taylor (1969) described a number of sections in the Kimmeridge Clay close to the southern boundary of the Ely district. The most famous section in the Kimmeridge Clay of East Anglia is that at Roslyn (or Roswell) Hole, Ely (Plate 4). This large complex of pits was begun in early Victorian times to provide embanking materials for the Great Ouse drainage system and has been almost continuously worked for that purpose to the present day. In the latter part of the 19th century these workings attracted numerous field parties from Cambridge and London, for not only was the Kimmeridge Clay at Roslyn richly fossiliferous, but the pits also exposed boulder clay and a large mass of Cretaceous rocks. The relationship of these latter to one another and to the Kimmeridge Clay was for many years a subject of controversy (see p.65 for details), and Seeley (1865a and b,1868), Fisher (1868), Bonney (1872a and b, 1875), Blake (1875) and Skertchly (1877) all entered into the discussion. Whitaker (1883), Roberts (1892), Whitaker and others (1891), McKenny Hughes (1884, 1894) and Rastall (1910) also discussed the geology of the pits or described exposures of Kimmeridge Clay within them.

The full local sequence of the Kimmeridge Clay was penetrated in two continuously cored boreholes drilled for oil shale exploration at Methwold Common and Decoy Farm, Southery (Pringle, 1923), and parts of the formation have been sampled in a number of shallow site-investigation and gravel-exploration boreholes. The Kimmeridge Clay sequence was also cored in the BGS borehole at Haddenham (Horton *in* Institute of Geological Sciences 1971; Medd 1979), close to the northern boundary of Sheet 188. The deep boreholes at Lakenheath and Soham showed the Kimmeridge Clay to be absent there, due to pre-Aptian erosion.

RWG

DETAILS

The sites of the exposures and boreholes described below are shown in Figure 17. In the following descriptions of the sections, the bed numbers (KC = Kimmeridge Clay) referred to are those of Figure 16.

Sutton – Wilburton

The junction of the Ampthill and Kimmeridge clays was formerly exposed in a large borrow-pit (Figure 13) at Bury Lane, Sutton [432 793]. The section was described by Forbes (1960, p.234) and the late Dr Arkell (Sedgwick Museum MS). In the following description the stratigraphical classification is based on a re-examination of specimens held in the Sedgwick Museum and BGS collections:

Bed No.		Thickness m
Clayey soil and subsoil overlying deeply weathered pale and medium grey clays with small calcareous pellets ('race')		1.4
KC 5:	Siltstone; medium grey with rare small angular phosphatic pebbles and common, well rounded, tabular cobbles (up to 25 x 20 x 5 cm) of partially phosphatised pale grey limestone that have numerous large, flask-shaped borings (*Gastrochaenolites*) on all their surfaces (Plate 3); shelly in part with common *Pachypictonia*, *Prorasenia* and *Rasenia*; wood fragments common	0.15
KC 2 to 4:	Mudstone; silty pale and medium grey weathering to yellowish grey; fauna mostly destroyed by weathering but including *Deltoideum delta*, *Gryphaea?*, locally common *Nanogyra*, pyritised inner whorls of perisphinctids and *Chondrites*	1.98
KC 1:	Mudstone; very pale grey with common burrowfills of medium grey silt; locally cemented to form a very pale grey argillaceous limestone; very shelly in part with abundant terebratulids and rare large, coarsely ribbed rhynchonellids including *Torquirhynchia* cf. *inconstans*; body-chamber infillings of ammonites common including *Pictonia?* and *Ringsteadia?*, some with grazing trails on their upper surfaces	0.46
Ampthill Clay		see p.30 for details

The presence of smooth-body-chambered perisphinctid ammonites, including forms assigned here to *Pictonia?*, together with common brachiopods indicates the presence of Bed KC 1 (*Pictonia baylei* Zone). *Pachypictonia*, *Prorasenia* and *Rasenia* are characteristic of the *Rasenia cymodoce* Zone and, where preserved in siltstone, indicate the presence of Bed KC 5 and/or KC 8. Evidence from elsewhere in the Ely district (see below) strongly suggests that Beds KC 5 and KC 8 are locally combined to form a single thick siltstone, but at Bury Lane the sparse fauna and thinness of the bed suggests that only Bed KC 5 is present. The basal bed of the Kimmeridge Clay in the BGS borehole at Haddenham was 1.1 m of weakly cemented, very shelly siltstone, with phosphatised *Pictonia?* in the lower part and abundant *Prorasenia* and *Rasenia* throughout. The lithology and fauna suggests that Beds KC 1, KC 5 and KC 8 are combined there and that the intervening beds (Beds KC 2 to KC 4, KC 6 and KC 7) have been removed by penecontemporaneous erosion (see p.31 for remainder of Haddenham Borehole section). The distinctive lithologies and faunas of Beds KC 5 and 8 can be traced southwards from Fenland as far as the type Kimmeridge Clay section of the Dorset coast.

Specimens of very fossiliferous siltstone that probably came from Bed KC 5/8 were obtained from newly cleared ditches at Grunty Fen Farm, Wilburton [488 769] (Holmes *in* Worssam and Taylor, 1969, p.22). Spath (1952, MS) described the ammonite fauna as indicating 'probably *baylei-cymodoce* zone or a new fauna from in between' and Casey (*in* Worssam and Taylor, 1969, p.22) attributed them to the *cymodoce* Zone. The ammonites include *Rasenia spp.* and rarer *Amoeboceras (Amoebites)*, some with encrustations and some with cream-coloured phosphatic infillings. The remaining fauna includes the bivalves *Modiolus?*, *Myophorella*, *Pholadomya*, *Pleuromya* cf. *alduini* Brongniart, *P.* cf. *uniformis* (J. Sowerby) and common oysters including large *Liostrea* and *Nanogyra virgula* (Defrance), some with encrusting serpulids. Until further work is done on the systematics of the genus *Rasenia*, this raseniid fauna can be assigned only tentatively to the group of *R. orbignyi* Spath non Tornquist; it

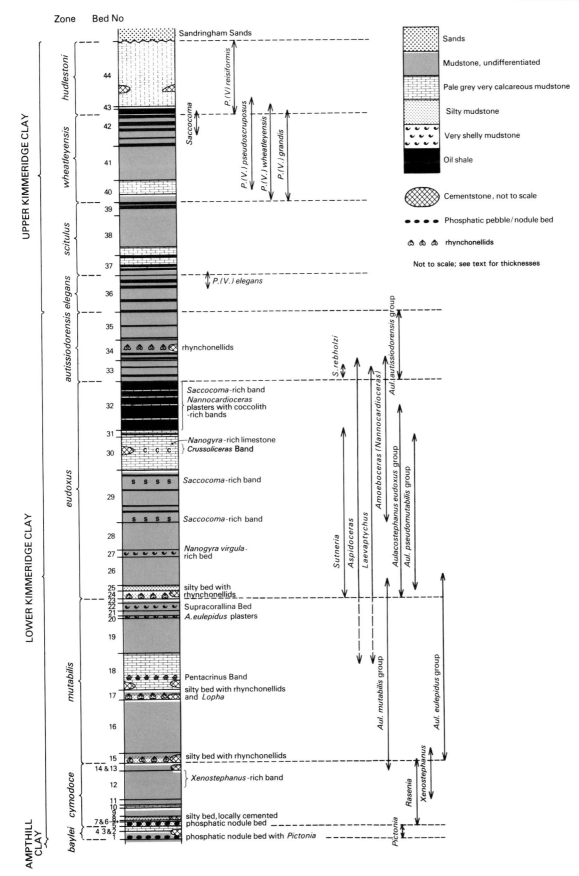

Figure 16 Generalised vertical section of the Kimmeridge Clay of the region

represents an early fauna of the *cymodoce* Zone (probably Bed KC 5). The presence of *Amoeboceras (Amoebites)*, common myids and oysters, together with the lithology suggests that Bed KC 8 also occurs.

Sandhill, Littleport

A similar lithology and fauna to that from Grunty Fen Farm were recorded in excavations for a coffer dam in the bed of the River Great Ouse at Sandhill, Littleport [577 865]. There, a shelly siltstone with angular black phosphatic pebbles, in part calcite-cemented, yielded a rich ammonite and bivalve fauna. Part of the shelly fauna is locally phosphatised. The ammonites are dominated by the genus *Rasenia*, and include macroconch (up to 350 mm) and microconch forms. Part of the raseniid fauna is comparable with that from Grunty Fen (*R. orbignyi* group), but much of it is undescribed. Casey (1960) compared some of these specimens with the genus *Pachypictonia* described from southern Germany. Species of *Amoeboceras*, including *A. Amoebites) cricki* (Salfeld), and the giant nautiloid *Paracenoceras giganteus* (d'Orbigny) are also present. The non-ammonite fauna includes the bivalves *Astarte?*, *Camptonectes*, *Corbulomima*, *Grammatodon* including *G. (Cosmetodon)*, myids including *Girardotia* aff. *compressa* (J. de C. Sowerby) and *Pleuromya* aff. *uniformis* (J. Sowerby), *Nuculoma*, oysters including *Nanogyra* and large '*Liostrea*', and encrusting *Placunopsis*, the gastropods *Pleurotomaria* and *Procerithium*, encrusting serpulids, and a fish tooth. Wood fragments are also present. A bored cementstone pebble with *Gastrochaenolites*, similar in lithology to that recorded from Bed KC 5 at Bury Lane, Sutton, was also collected.

Figure 17   List of exposures referred to in the text

1	Bury Lane, Sutton
2	Grunty Fen, Wilburton
3	Sandhill, Littleport
4	Haddenham End brick pits
5	Fordham Road and Ferry Bridge, Stretham
6	Roslyn Hole, Ely
7	Chettisham Station
8	California brick pit, Downham
9	Littleport brick pits
10	Black Horse Farm, Brandon Creek
11	Engine drain, Southery
12	Southery by-pass

Haddenham

Kimmeridge Clay was formerly worked for brickmaking in extensive pits on both sides of the railway at Haddenham End. Museum collections labelled 'Haddenham', 'Haddenham End' and 'Haddenham Station' all probably came from this complex of workings. Roberts (1892, pp.63–65) described two pits [468 762 and 467 761] on the south side of Haddenham railway station and a third to the north-west of the station [probably 462 767]. Kitchin and Pringle (1926, MS) subsequently collected additional fauna from one of these pits but did not provide any section that could be correlated with that described by Roberts. Their collection includes richly

Plate 4   Kimmeridge Clay at Roslyn Hole [552 806], Ely. This complex of pits has been used almost continuously since Victorian times to provide clay to repair the banks of local Fenland drains and rivers. The sections are now badly degraded: they show fossiliferous shales, mudstones and calcareous mudstones in which a line of cementstone doggers (arrowed) in Bed KC 30 forms a prominent marker band (A13722)

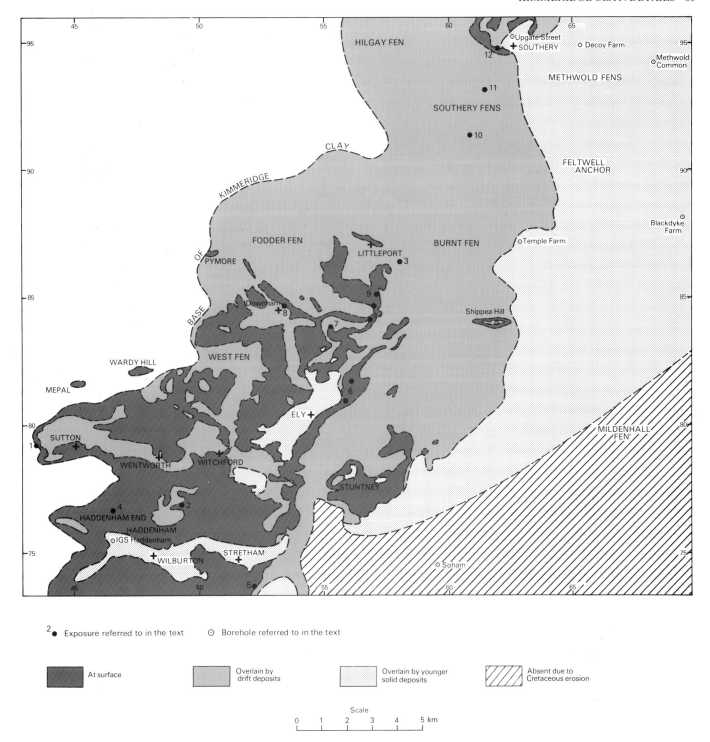

2 ● Exposure referred to in the text   ⊙ Borehole referred to in the text

■ At surface	Overlain by drift deposits
Overlain by younger solid deposits	Absent due to Cretaceous erosion

Scale
0 1 2 3 4 5 km

Figure 17 Sites of Kimmeridge Clay boreholes and exposures referred to in the text

fossiliferous, pale and very pale grey calcareous mudstones and a bed of septarian cementstones. The mudstones contain *Amoeboceras*, *Aulacostephanus (Aulacostephanites)* cf. *eulepidus* (Schneid), *A. (Aulacostephanoides)* cf. *linealis* (Quenstedt) and *Aspidoceras*, and the cementstones yielded serpulid- and oyster-encrusted specimens of *Aulacostephanus (Aulacostephanoides) mutabilis* (J. Sowerby), *A. (Aulacostephanoides)* cf. *sosvaensis* (Sasonov) and '*Subdichotomoceras*'. A rich bivalve fauna, including *Astarte*, *Corbulomima*, *Grammatodon*

(*Cosmetodon?*), *Isocyprina*, *Protocardia?*, *Thracia* and oysters, is also present together with columnals of the crinoid *Pentacrinus*, and the gastropod *Dicroloma*. The ammonite assemblage indicates the *Aulacostephanus mutabilis* Zone; the occurrence of *Aspidoceras* suggests the upper part of the zone or the basal part of the *Aulacostephanus eudoxus* Zone. The faunal and lithological association is that of Bed KC 18, and includes a widespread marker bed, the Pentacrinus Band. Roberts (1892) recorded a cementstone band in the pits on

Figure 18 Geological sketch map of Roslyn Hole (Roswell Pits), Ely

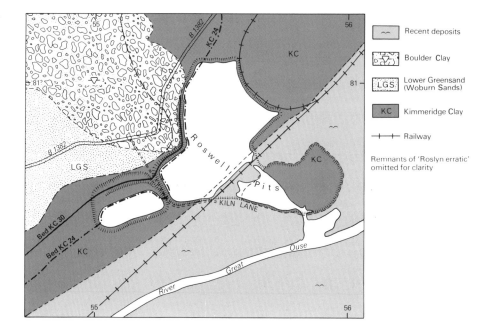

the south side of the railway, and material collected by him (now in the Sedgwick Museum) can be lithologically and faunally matched with the material described above (Bed KC 18). Ziegler (1962) figured some *Aulacostephanus* from this locality. Robert's third pit, on the northern side of the railway, yielded abundant *Deltoideum delta*, some with encrusting *Lopha*; *Nanogyra*, terebratulid brachiopods, *Rasenia?* and a phosphatic pebble bed, and seems likely to have penetrated the *cymodoce* Zone.

Stretham

A highly fossiliferous Kimmeridge Clay section in a large borrow-pit at Fordham Road, Stretham [516 743] on the west bank of the River Cam in the northern part of the Cambridge district has been described by Forbes (1960, p.235), who assigned the sequence to the *baylei*, *cymodoce*, and *mutabilis* (= *pseudomutabilis*) zones, and by Worssam and Taylor (1969, pp.22–23), who assigned it to at least the *mutabilis* Zone. The lowest part of the section, Beds 1 to 3 of Forbes (1960), consists of pale and very pale grey calcareous clays with a band of septarian cementstones. The cementstones have yielded smooth body-chamber fragments of *Aulacostephanus* (*Aulacostephanoides*) of the *mutabilis* group (misidentified by earlier authors as *Pictonia*), *A.* (*Aulacostephanites*) *eulepidus*, *A.* (*Aulacostephanoides*) *linealis*, *Laevaptychus*, *Isocyprina* (*Venericyprina*) cf. *argillacea* Casey, large oysters, serpulids and a plesiosaur tooth. The adjacent clays contain *A. eulepidus*, *Isocyprina* (*Venericyprina*) *elongata* (Blake), *Lima?* *concentrica* (J. de C. Sowerby), *Protocardia* sp. indet., *Thracia depressa* (J. de C. Sowerby), nuculoids, oysters including *Nanogyra*, *Proceritium* and *Pentacrinus* columnals. The assemblage and lithology is that of Bed KC 18 and includes the Pentacrinus Band. The presence of common *Laevaptychus* in the upper part of the section suggests that the *eudoxus* Zone is also present. Bones of a giant pliosaur were collected here at the time of working: these were subsequently re-assembled by Tarlo (1958, 1959a) and described as a new genus, *Stretosaurus*.

A similar section was proved in a nearby borrow pit at Stretham Ferry Bridge [502 723] (Worssam and Taylor, 1969, p.22) where pale grey clays with a nodular cementstone band yielded *Aulacostephanus eulepidus*, *A. mutabilis* group, *Aspidoceras* fragments, *Corbulomima?*, *Isocyprina?*, *Lingula?* and *Pentacrinus* columnals.

Ely

The most famous section in the Kimmeridge Clay in East Anglia is that at Roslyn Hole, Ely [555 808], where clay has been worked for more than a century to provide embanking materials. Blake (1875, pp.217–222), Roberts (1892, pp.65–70), and Whitaker and others (1891, pp.18–19) described extensive faunas from this pit. The large faunal collections, labelled 'Roslyn', 'Roswell' or simply 'Ely', in the BGS and the Sedgwick Museum collections came from this locality. The pits referred to by the 19th century authors lay on both sides of the railway (Figure 18) and are now flooded. A new pit was opened by the River Authority in the 1930s, and this still exposes a similar Kimmeridge Clay section to that described by Roberts (1892, pp.65–69). The following section was measured during the present study in the west face of the modern pit [552 807]:

Woburn Sands		see p.49 for details

Kimmeridge Clay

Bed No.		Thickness m

Aulacostephanus eudoxus Zone

KC 31 & 32:	Oil shale and pale grey calcareous mudstone interbedded in 5 to 10 cm-thick units; very shelly in part with '*Lucina*', *Protocardia*, abundant *Amoeboceras* (*Nannocardioceras*), *Aulacostephanus*, *Discinisca*, *Dicroloma*, fish debris and shell dust	2.45
	Mudstone, pale and medium grey, moderately shelly with fauna as bed above; serpulids locally common; *Lingula* and coalified wood fragments common; 8 cm-thick oil shale in middle part of bed	1.33
	Oil shale, very shelly, fauna as bed above but with *Aulacostephanus* and faecal pellets especially common	0.06
KC 30:	Mudstone, pale and very pale grey with bioturbation at several levels picked out by darker infillings; sparsely shelly in upper part but with abundant serpulids,	

Bed No.		Thickness m
	Nanogyra virgula (Defrance) and other small oysters in lower part	1.40
	Cementstone, prominent band of closely spaced (1.0 to 1.2 m apart) doggers set in pale grey calcareous mudstone; flattened spheroidal and ellipsoidal shapes mostly 35 to 40 cm (up to 47 cm) thick and 1.0 to 1.2 m in diameter; septarian cracks in cores partially infilled by sparry calcite; 0.15 to 0.20 cm thick band of weakly calcite-cemented mudstone occurs between doggers at the level of their horizontal axis; sparsely shelly but with relatively common *Nanogyra virgula* and other oysters, and rare *Crussoliceras*	0.40
	Mudstone, pale and medium grey sparsely shelly	0.36
	Oyster lumachelle composed almost entirely of *N. virgula* (Virgula Limestone)	0.07
	Mudstone, dark grey, sparsely shelly to shelly with *Amoeboceras (Amoebites)*, serpulids and rotted bivalves and other ammonites; layer with abundant *N. virgula* at base	0.40
	Mudstone, thinly interbedded pale, medium and dark grey; mostly sparsely shelly but with some shelly bands; fauna as bed above	1.75
KC 29:	Mudstone, dark grey with thin seams of muddy oil shale; poorly exposed; passing down into	1.10
	Mudstone, medium grey, silty, sparsely shelly passing down into	0.15
	Siltstone, medium grey, weakly calcite-cemented with widely spaced cementstone doggers (10m to 15m apart) of similar size to those in the higher bed; shelly with *Aulacostephanus* of the *eudoxus* group; *Aspidoceras, Laevaptychus*, poorly preserved bivalves and fish debris; interburrowed junction with	0.17
	Mudstone, pale grey becoming darker with depth; shelly in lower part; interburrowed junction with	0.40
	Oil shale, brownish grey, muddy, deeply weathered	0.10
	Siltstone, medium to pale grey, with widely spaced (more than 30 m apart) cementstone doggers; locally shelly in lower part with common *Aspidoceras* and *Laevaptychus*; interburrowed junction with	0.12
	Mudstone, pale grey, sparsely shelly	0.20
	Mudstone, dark grey, very shelly with *Aspidoceras, Laevaptychus* and *Nanogyra virgula*	0.09
	Mudstone, pale and medium grey, sparsely shelly	0.65
	Mudstone, dark grey, shelly; fauna of bivalves and ammonites rotted by weathering but *Laevaptychus* occurring loose in spoil	0.45
	Mudstone, poorly exposed but mostly medium and pale grey; sparsely shelly but with some rotted shelly bands	c.1.5

KC 25 to ?28:	Mudstone, pale grey, moderately shelly with bivalves and ammonites, including *Aspidoceras*, in rotted pyritic preservation	0.30
	Mudstone, medium grey, fissile, shelly; some pyritic preservation of bivalves and ammonites; possible phosphatised burrowfills	0.15
	Mudstone, pale grey, sparsely shelly, becoming darker in lower part and passing down into	0.55
KC 24:	Siltstone, medium grey, pyritic but deeply weathered to form a prominent yellow band with sulphur-coated surfaces; very shelly with fauna concentrated in lower part; *Aspidoceras* and *Aulacostephanus eulepidus* common, *Astarte supracorallina* d'Orbigny, oysters and fish debris also present	0.15
Aulacostephanus mutabilis Zone		
KC 21:	Mudstone, medium grey, sparsely shelly	0.50
KC 20:	Mudstone, faintly brownish grey (kerogen-rich), fissile; very shelly with abundant *Aulacostephanus eulepidus*, possible *A. mutabilis*, 'Astarte' supracorallina and faecal pellets	0.15
KC 19:	Mudstone, medium and pale grey, sparsely shelly with concentrations of large, smooth body-chambered ammonites (*A. mutabilis?*) at several levels	0.90 *seen*

The lowest horizon exposed in the pit is a pale grey deeply weathered mudstone (Bed KC 18) that occurs 1 to 2 m below the lowest beds of the main section. Part of the sequence described above is also exposed in the north-western corner of the pit [554 810] where the two siltstone horizons in Bed KC 29 are exposed, the upper one with cementstone doggers at 2 to 4 m spacings. Roberts (1892, p.66) recorded about 13 m of beds on the northern side of the pit [probably 555 810] which appear, from his lithological description and faunal collection, to include representatives of Beds KC 18 to KC 32. Skertchly (*in* Whitaker and others, 1891, p.16) measured about 8m of these same beds in the south-eastern corner of the pit [probably 555 806]. Roberts (1892, p.69) recorded a second pit on the east side of the railway [probably 557 806] at a similar stratigraphical level. All these sections now lie within a single flooded complex of pits.

All the ammonites from Ely in the BGS and the Sedgwick Museum collections are indicative of the *mutabilis* or *eudoxus* zones and could have come from the present-day exposure. This is surprising, because the depth (up to 12 m) and large areal extent of the workings, combined with the low westerly dip, suggest that the whole of the Lower Kimmeridge Clay might formerly have been exposed. The non-ammonite fauna, with the notable exceptions of the brachiopod *Torquirhynchia inconstans* (J. Sowerby) and the bivalve *Deltoideum delta*, could also have come from Beds KC 18 to 32. The presence of *T. inconstans* at Roslyn Hole was noted by Bonney (1875), Roberts (1892) and McKenny Hughes (1894), but there is no record of it ever having been collected *in situ*. The last of these authors described an excavation made below water level and 'not available at the time of Roberts' measurements' from which 'amongst other finds there were some remarkably large *R. inconstans*'. The section was not described and it seems likely, in the absence of an accompanying ammonite fauna, that these *Torquirhynchia* came either from the drift or from an erratic mass of Kimmeridge Clay within the drift (see p.65).

Other fauna in the Sedgwick Museum collection include bivalves, belemnites, inarticulate brachiopods, echinoid spines, fish including a well-preserved specimen attributed by Woodward

(1890) to *Eurycormus*, serpulids, and plant and reptilian debris including a relatively complete specimen of *Pliosaurus* identified by Tarlo (1959b) as *P. brachyspondylus* (Owen). The hind limb bones of this last named specimen were damaged by an osteochondritic disease (Wells, 1964).

Chettisham – Downham

Part of the Roslyn Hole sequence was formerly exposed in a railway cutting (widened at its eastern end to provide fill for the Ely to March railway where it crosses the Bedford rivers) near Chettisham Station [549 837]. There, Roberts (1892, p.70) recorded about 3 m of clay rich in *Nanogyra virgula* and with lines of septarian cementstones at the top and bottom of the section. The fauna includes *Aspidoceras*, together with *Laevaptychus*, *Amoeboceras* (*Amoebites*) and *Crussoliceras*, common *N. virgula* and common serpulids, all in either cementstone or pale grey calcareous clay preservation. This fauna and lithology are characteristic of Bed KC 30. However, other material labelled 'Chettisham' in Roberts' collection includes *Pictonia?* and *Rasenia* preserved in a shelly siltstone (basal bed of the Kimmeridge Clay) and a specimen of *Torquirhynchia inconstans*. These were presumably obtained from erratic blocks of siltstone within the Chalky-Jurassic till that crops out in the western part of the same cutting. Skertchly (*in* Whitaker and others, 1891, p.70) noted that the till in this area was so rich in Kimmeridge Clay as to be difficult to distinguish from *in situ* Kimmeridge Clay.

Kimmeridge Clay was formerly worked for brickmaking at California, Downham [531 844]. The section has now degraded to a pale grey calcareous clay: no description or collection appears to have survived from this locality.

Littleport brick-pits

Roberts (1892, pp.70–72) recorded three brick-pits close to the road from Ely to Littleport. The first, at Brick Hill (near Pyper's Hill), Littleport [565 841], worked up to 6 m of Kimmeridge Clay beneath a thin covering of Recent deposits. The second, 800 m due north of the first [565 849], contained up to 4 m of clay with two limestone bands (Roberts, 1892, p.71). The third pit, in the south-eastern part of Littleport Fields [566 854], worked up to 8 m of clay with three horizons of cementstone. Roberts (1892) noted that the dip in the second pit was gently northwards, and in the third pit north-westwards at about 5°. He therefore concluded that the most southerly pit (the first) exposed the oldest beds and the most northerly pit the youngest.

The two southerly pits have been backfilled: the northern pit now shows about 2 m of deeply weathered pale grey clay in its lowest part. The faunal evidence is poor and the few fossils that have survived from these localities are labelled 'Littleport'. The only determinable ammonite is *Aulacostephanus eulepidus*; the bivalves include arcids, '*Lucina*', nuculoids, *Thracia*, and *Trigonia*. Where visible, the matrix material is a pale grey calcareous clay. This lithology together with the cementstones and the fauna makes it likely that at least one of the pits exposed Bed KC 18. The occurrence of abundant *Astarte supracorallina* in the upper part of the most northerly pit (Roberts, 1892, p.72) suggests the presence of Bed KC 22; it seems likely that the underlying beds there, which included two cementstone bands, are part of Bed KC 18. These lower beds also yielded specimens attributed to '*Rhynchonella*' *inconstans* (Sedgwick Museum collection).

Rhynchonellids have been recorded *in situ* in the Kimmeridge Clay at only two localities in the Ely district; at Bury Lane (Forbes, 1960) rare, more or less crushed, coarsely ribbed rhynchonellids were found in the terebratulid-rich limestone that marks the base of the Kimmeridge Clay (see p.36 for details) and at Roslyn Hole, Roberts (1892) recorded rhynchonellids *in situ* in his Beds 10, 11 and 12 which equate with the lower siltstone and immediately

underlying clays of Bed KC 29 (see p.41). Rhynchonellids from Chettisham and Littleport (Roberts, 1892), Stretham (Forbes, 1960), and most of those from Bury Lane (Forbes, 1960) and Roslyn Hole (McKenny Hughes, 1894), were collected loose. All these specimens except the *in situ* material from Roslyn Hole (identified by Roberts (1892, p.68) as *R. pinguis?* Roemer) have been attributed in the past to *Torquirhynchia* [formerly *Rhynchonella* or *Rhactorhynchia*] *inconstans* (J. Sowerby). Although comparable in size to mature *T. inconstans*, most are crushed and appear to have been markedly less globose and generally more coarsely ribbed than specimens of this species from the basal bed of the Kimmeridge Clay in Dorset. In addition, in those specimens which are sufficiently uncrushed for their original shapes to be determined, the commissure does not show the marked assymmetry characteristic of true *inconstans*; in general aspect they bear more resemblance to the species *sutherlandi* Davidson.

A few of the smaller specimens attributed to *inconstans* from Bury Lane (collected loose but presumed to come from Bed KC 1 from their adhering matrix), Roslyn Hole and Stretham resemble small *T. inconstans* from Dorset and are assigned herein to *T.* cf. *inconstans*. With the exception of those from Bury Lane, their stratigraphical position is unknown.

The Kimmeridgian rhynchonellids from the Ely district thus include three groups which have been compared with the species *pinguis*, *sutherlandi* and *inconstans*. More definite determination and discussion of the stratigraphical significance of these forms must await their more detailed systematic description. It is noteworthy, however, that the basal bed of the Kimmeridge Clay in both the Ely district and the Dorset type area appears to be characterised by large, coarsely ribbed rhynchonellids.

Brandon Creek

Specimens collected in 1951 from a borrow-pit [610 908] near Bank Farm, Brandon Creek consist of oil shale, pale grey mudstone and cementstone. The fauna includes *Aulacostephanus* (*Aulacostephanoceras*) cf. *eudoxus* (d'Orbigny), *A.* (*A.*) cf. *pseudomutabilis* (de Loriol), *Aspidoceras*, *Laevaptychus*, *Grammatodon*, *Isocyprina*, *Nanogyra virgula*, *Pinna*, *Protocardia*, oysters, bivalve debris and spat, *Lingula?*, fish fragments and a plesiosaur tooth. This assemblage, taken in conjunction with the lithologies, indicates the presence of Beds KC 29 and 30.

Southery

Spoil from the foundations of a new pumping station at the junction of the Engine Drain and the Great Ouse [6128 9318] included cementstone doggers and mudstones with fragments of *Aulacostephanus* and abundant *Nanogyra virgula* (probably Bed KC 30). The Recent deposits of this part of Southery Fens are thin, and many of the deeper drains lying on either side of the Great Ouse have cut into the weathered upper surface of the Kimmeridge Clay.

Rose (1835, pp.174–175) recorded a brick-pit at Southery [probably 617 958] and noted that it exposed 4 m of 'brickearth', underlain in the floor of the pit by a 5 to 7 cm seam of inflammable shale. The bulk of the pit is in pale grey clays with cementstone doggers (Bed KC 44). It was presumably this calcareous clay that gave rise to the yellow stock brick that is much in evidence in the Victorian and Edwardian parts of the village. The underlying oil shales are those of Bed KC 42.

Excavations for the Southery Bypass [616 946] showed pale grey calcareous clays (Bed KC 44) overlain by chalky boulder clay rich in local Upper Jurassic and Lower Cretaceous debris. Fitton (1836, p.316) collected two characteristic Lower Kimmeridge Clay fossils at Southery, the large flat oyster *Deltoideum delta*, and *Laevaptychus*. *Deltoideum delta* is known only from the Ampthill Clay and the basal beds of the Kimmeridge Clay and *Laevaptychus* only from the *mutabilis* to *autissiodorensis* zones (Beds KC 18 to 33).

The oldest Kimmeridge Clay at outcrop on the 'island' of Southery is in Bed KC 42 and could not be the source of these fossils. However, both fossils have thick calcitic shells and are common in the boulder clay of west Norfolk.

Boreholes

Kimmeridge Clay has been proved in a number of cored boreholes in and close to the district. According to Pringle (1923, pp.128–132), the Methwold Common and Severals House boreholes proved 36.9 and 38.1 m of Kimmeridge Clay respectively. These interpretations were subsequently revised by Arkell (1937, p.82) to 43.6 m and 44.8m. A third borehole, at Decoy Farm, Southery [6485 9482], penetrated the top 30.6 m of the formation (Pringle, 1923, p.127). Pringle (1923, p.134) believed the full sequence of Kimmeridgian zones was present in both the Methwold Common and Severals House boreholes and attributed this unusually thin Kimmeridge Clay, as compared to the Dorset type section where Arkell (1947, p. 64) estimated the formation to be more than 500 m thick, to an overall attenuation resulting from slow deposition, rather than to the overstep of the Sandringham Sands. This interpretation has subsequently been shown to be partially correct (Gallois, 1973): the Kimmeridge Clay in Norfolk is considerably attenuated in comparison with the equivalent beds in Dorset, but a large part of the Norfolk sequence has also been lost by erosion in the late Jurassic. The youngest Kimmeridge Clay in all three boreholes belongs to the *Pectinatites hudlestoni* Zone (Bed KC 44) and the highest zones (c. 180 m of strata in Dorset) are not represented. Few specimens have survived from these boreholes but those that have include a number of distinctive lithological and faunal associations. Silty mudstone with phosphatised *Amoeboceras* and *Rasenia?* (Bed KC 5/8) was proved in the Severals House and Methwold Common boreholes. Pale grey mudstone with abundant '*Astarte*' *supracorallina* (Bed KC 22), siltstone with common *Aulacostephanus eulepidus* (Bed KC 24), and pale grey mudstone with large *Nanogyra virgula* (Bed KC 30) was proved at Severals House. Oil shales with abundant *Nannocardioceras* (Bed KC 32) and oil shales with *Saccocoma* (Bed KC 42) overlain by pale grey mudstone (Bed KC 44), were proved in all three boreholes.

The junction of the Kimmeridge Clay and the Sandringham Sands was proved in shallow boreholes at Upgate Street [6208 9483] and Decoy Farm [6490 9463], Southery. The youngest Kimmeridge Clay at both localities was a softened (due to ground water leaching) pale grey mudstone (Bed KC 44).

Kimmeridge Clay was proved at relatively shallow depths (1.8 to 10.8 m below ground level) in a large number of site-investigation boreholes drilled between Southery and Ely (for the River Great Ouse improvement) and between Southery and Brandon Creek (A10 trunk road improvement).

Two boreholes [6292 8685 and 6285 8726] at Temple Farm, Little Ouse proved Sandringham Sands resting on medium and pale grey mudstones with abundant *Nanogyra virgula*, *Amoeboceras (Amoebites)* and *Aulacostephanus* (probably Bed KC 30). A site-investigation borehole [6916 8817] at Blackdyke Farm, Hockwold proved a similar sequence of Sandringham Sands on Bed KC 30. There the Kimmeridge Clay fauna included abundant serpulids, common *Lingula*, large *Nanogyra virgula*, *Amoeboceras (Amoebites)* and *Aspidoceras?*.

Kimmeridge Clay (probably Beds KC 22 to 26) was proved beneath thin drift deposits in a borehole [6158 8384] at Shippea Hill Farm, Shippea Hill. The basal bed of the *eudoxus* Zone (Bed KC 24) with a rich shelly fauna including *Amoeboceras (Amoebites)*, *Aspidoceras*, *Aulacostephanus*, *Laevaptychus* and abundant bivalves, was encountered near the bottom of the borehole, but was overlain by deeply weathered clays.

In a borehole [5444 8053] at Bray's Lane Ely, the Woburn Sands rested on oil shales with *Nannocardioceras* (Bed KC 32), as at the nearby exposure at Roslyn Hole.

The continuously cored BGS borehole at Haddenham proved the total thickness of the formation there to be 21.84 m (Horton *in* Institute of Geological Sciences, 1971, p.104). The basal bed of the formation was lithologically and faunally similar to that proved at outcrop elsewhere in the district and consists of 1.1 m of siltstone resting on an irregular burrowed surface of Ampthill Clay. This bed probably includes parts of Beds KC 1, 5, and 8 (see p.00). Most of the more distinctive lithologies between Beds KC 8 and 24 can be recognised in the borehole, although the sequence is thinner than elsewhere in Fenland. Bed KC 24 is well developed as a siltstone with *Aspidoceras*, *Aulacostephanus eulepidus* and rhynchonellid brachiopods. Above Bed KC 24, the sequence can be matched in detail with that proved at Roslyn Hole, Ely (see p.00), and includes the oil shales in Bed KC 29, the Virgula Limestone (Bed KC 30) and the oil shales in Bed KC 32: the Woburn Sands rest unconformably on Bed KC 32, as at Ely.

Kimmeridge Clay has also been proved beneath younger deposits in boreholes drilled for water at Corkway Drove [6763 8972], and Shrubhill Farm [6618 8806], Feltwell Anchor, Forty Farm [6522 7922] and Cooks Drove Farm [6469 7838], Mildenhall, and at Bradford Farm [5669 7724], Stuntney, but no detail has been recorded.   BMC, RWG

SANDRINGHAM SANDS

The Sandringham Sands of Norfolk consist of clean white, fine-grained quartz sands, green glauconitic sands and grey clayey sands with subordinate thin beds of siderite and plant-rich mudstone. In south-west Norfolk, the formation is progressively overstepped in a southerly direction by the Carstone (Cretaceous); only the lowest member of the Sandringham Sands, the Roxham Beds, is preserved in the Ely district.

The Roxham Beds consist of up to 6 m of grey and green, bioturbated fine-grained sands, in part clayey or glauconitic and with a few pyrite concretions. The formation rests unconformably on the Kimmeridge Clay. Its base is marked by a pebble bed containing water-worn, phosphatised casts of bivalves and pavlovid ammonites derived from the Kimmeridge Clay. The basal 0.5 to 1.0 m consists of a densely calcite- and, in part, pyrite-cemented, highly shelly sandstone containing the large perisphinctid ammonite *Paracraspedites*, the large brachiopod *Rouillieria ovoides* (J. Sowerby) (formerly *Terebratula rex*), and trigoniid and pectinid bivalves. This distinctive bed is usually decalcified at outcrop and its fauna destroyed, but large unweathered boulders of it are common as erratics in the Chalky-Jurassic till (see p.65) of East Anglia. Such boulders have been recorded at Southery and Ely.

Casey (1962; 1967; 1973) described the ammonites from this basal bed and compared them with those from the Portland Stone of southern England and from Upper Jurassic strata of the Russian Platform. He proposed (Casey, 1973) a zonal classification for this stratigraphical level in which the Roxham Beds fall within the Zone of *Paracraspedites oppressus* (Middle Volgian). He suggested (1973) that this zone is represented in the topmost part of the Portland Stone where it is characterised by titanitid ammonite faunas. William Smith noted in the margin of his geological map of Norfolk (1819) that the 'sands between the Oaktree Clay [Kimmeridge Clay] and the Golt Brickearth [Gault]' included locally at their base a bed of 'Portland Stone'. Later workers, notably Rose (1835-36, 1859, 1862), Fitton (1836), Teall (1875) and Whitaker and others (1893), considered the

Sandringham Sands to be wholly Cretaceous in age, so that more than 140 years elapsed between Smith's perceptive observation and its verification by Casey.

In the Ely district, the remainder of the Roxham Beds consists of apparently unfossiliferous soft sands. Elsewhere in Norfolk, these beds have yielded pyritised *Paracraspedites* which show the Roxham Beds to be wholly Jurassic in age. Patches of pebbly glauconitic sand containing the Upper Volgian ammonite *Subcraspedites* have been recorded as erratics in the boulder clay at Southery. These are derived from a younger member of the Sandringham Sands, the Runcton Beds, which crops out in the Wisbech and King's Lynn districts.

The only outcrop of Roxham Beds in the Ely district is on the crest and eastern flanks of the 'island' of Southery where the sands give rise to a grey and yellow sandy subsoil and were formerly worked for building purposes. The basal sandstone has not been worked because it appears to be everywhere deeply weathered at outcrop and was probably extensively decalcified during the Pleistocene. The formation has been proved in boreholes at Southery, at Methwold Fen, and at Little Ouse, close to its southern limit. The Roxham Beds are overstepped by the Carstone in the northern part of the district and by the Woburn Sands at Little Ouse (see p.46). These relationships are commonly difficult to determine precisely because of the lithological similarity of all three formations, and because the younger formations have incorporated much debris from the older.

DETAILS

Southery

Boreholes at Upgate Street [6208 9483] and Decoy Farm [6490 9463], Southery, proved 4.95 and 6.1 m of Roxham Beds respectively, underlain by Kimmeridge Clay and overlain by Carstone at Decoy Farm. The Roxham Beds in both boreholes consist of fine- and very fine-grained grey and greenish grey glauconitic sands with beds of dark grey carbonaceous clay: at their base, a pebble bed composed largely of vein quartz, lydite and phosphate pebbles rests on an irregular burrowed surface of Kimmeridge Clay.

The Decoy Farm [6485 9482], Methwold Common [678 941] and Severals House [6921 9639] oil-shale boreholes probably proved sequences similar to that at Southery, although Roxham Beds were not recorded by Pringle (1923, pp.126–133). At Methwold Common and Severals House, the basal pebble bed of the Roxham Beds was calcite- and pyrite-cemented as recorded elsewhere in Norfolk below the depth of weathering.

Little Ouse

About 2.5 m of Roxham Beds, consisting of fine-grained glauconitic sand with thin beds of dark grey carbonaceous clay, were proved beneath the Woburn Sands and above the Kimmeridge Clay in two boreholes at Temple Farm [6292 8685 and 6285 8726], Little Ouse. The basal pebble bed was again present, but decalcified. This is the most southerly known occurrence of the Roxham Beds. Up to 1m of silt and fine-grained sand may also have been penetrated between the Woburn Sands and Kimmeridge Clay in a borehole at Blackdyke Farm [6916 8817], Hockwold, but most of the core at this level was washed out during drilling.   RWG.

CHAPTER 5

Cretaceous

Cretaceous rocks underlie much of the eastern part of the Ely district but are mostly covered by the Recent deposits of Fenland. The older Cretaceous rocks, the Woburn Sands and Carstone (loose sands and clayey sands) and the Gault (soft mudstones), are deeply weathered everywhere at outcrop. In the extreme east of the district the Chalk forms an escarpment that marks the eastern limit of Fenland: even there, the relief is low and exposures are poor.

William Smith, in his geological maps of Cambridgeshire (1819) and Norfolk (1819), was the first to delineate accurately the Cretaceous rocks of the district. He recognised three formations, the 'Sand beneath the Golt', the 'Golt Brick Earth' and the 'Chalk' (Table 4). The last two of these have remained essentially unchanged, as the Gault and Chalk on the modern map. The deposits included in the 'Sand beneath the Golt' include several modern formations and it has only recently been understood how these relate to Cretaceous formations elsewhere in southern England.

William Smith appears to have first used the term 'Greensand', at some time between 1800 and 1812, to describe the glauconitic sands lying between the Gault and the Chalk in southern England. Some subsequent authors confused this bed with lithologically similar sands underlying the Gault in Kent and Sussex, hence William Smith's use of the phrase 'Sand beneath the Golt' for the beds between the Kimmeridge Clay and the Gault in the Ely district. After much discussion (see Jukes-Browne, 1900, pp.15–26 for summary) these last named sands were termed the Lower Greensand by Webster (1824, quoted in Jukes-Browne, 1900), and the stratigraphically higher sands became the Upper Greensand.

Rose (1862, pp.234–236) noted that the beds between the Kimmeridge Clay and the Gault (in south-west Norfolk) and between the Kimmeridge Clay and the Red Chalk (in north-west Norfolk) could be divided into three: 'loose white sand'; a ferruginous sandstone known locally as 'Carstone' (also known as Carr Stone = Fen Stone); and a 'breccia' (the Carstone of Hunstanton cliffs). The lowest of these divisions later became the Sandringham Sands, and the upper two became the Carstone of modern literature. The whole sequence was still considered at that time to be the lateral equivalent of the Lower Greensand of southern England and was shown as such on the Geological Survey maps (Sheet 65, 1886; Sheet 69, 1885). Thus, Whitaker and others (1891, 1893) interpretation of the 'Lower Greensand' in the Ely and adjacent districts encompassed beds now known to range from late Jurassic (Volgian) to mid-Cretaceous (Albian) in age (Table 4). However, by the time the memoir for Sheet 69 was published it had become clear that the Sandringham Sands were older than the Lower Greensand (Lamplugh *in* Whitaker and Jukes-Browne, 1899, p.22).

Table 4 Nomenclature and zonal scheme for the late Jurassic and Cretaceous rocks of the Ely district

Wm Smith, 1819	Whitaker and others, 1891 and 1893	This memoir		Zone	Stage	System
Chalk	Lower Chalk	Lower Chalk		*Acanthoceras rhotomagense* *Mantelliceras mantelli*	Cenomanian (pars)	
Golt Brick Earth	Gault	Gault		*Stoliczkaia dispar* *Mortoniceras inflatum* *Euhoplites lautus* *Euhoplites loricatus* *Hoplites dentatus*	Albian	Cretaceous
Sand beneath the Golt	Lower Greensand	Lower Greensand	Carstone	? *Douvilleiceras mammillatum* ? *Leymeriella tardefurcata*		
			Woburn Sands	? *Parahoplites nutfieldiensis* ? *Cheloniceras martinioides*	Aptian (pars)	
		Sandringham Sands (pars)		*Paracraspedites oppressus*	Volgian (pars)	
Oak Tree Clay	Kimmeridge Clay	Kimmeridge Clay		*Pictonia baylei* to *Pectinatites hudlestoni*	Kimmeridgian	Jurassic

? denotes that identification of the zone is uncertain.   ⌁⌁⌁   unconformity

In the Ely district, the true Lower Greensand is made up of two distinctive deposits of differing ages. The older of these consists of medium- and coarse-grained glauconitic sands that form a continuous and relatively uniform deposit of late Aptian (probably *Parahoplites nutfieldiensis* Zone) age that can be traced southwards at outcrop from the Little Ouse River to the Leighton Buzzard area of Buckinghamshire. In the present account this deposit is referred to as the Woburn Sands, following the suggestion by Cameron (1892) endorsed by Hancock (1972) and Rawson and others (1978). In the Cambridge, Huntingdon and Biggleswade districts the Woburn Sands were referred to as the 'Lower Greensand'.

The younger deposit is of early Albian (probably *Leymeriella tardefurcata* and/or *Douvilleiceras mammillatum* zone) age and consists of ferruginous sands and clayey sands. It can be traced northwards at outcrop from the Little Ouse River to Hunstanton, Norfolk, and has been referred to as the Carstone since the time of Rose (1862).

Throughout large areas of southern England the Lower Greensand rests with marked unconformity on Jurassic and Lower Cretaceous strata and the position of its lower boundary is rarely in doubt. The upper boundary is more problematical. In the Weald of Kent and Sussex the bulk of the Lower Greensand falls within the Aptian Stage but part of the highest subdivision, the Folkestone Beds, is largely early Albian (*tardefurcata* and *mammillatum* zone) in age. The *mammillatum* Zone part of the Folkestone Beds has been interpreted by some authors as the basal bed of the Gault because it commonly consists of sandy clays with phosphatic pebble beds. Casey (1961b) has attempted to clarify the classification by drawing the upper boundary of the Lower Greensand at a change from predominantly arenaceous to argillaceous sedimentation that coincides with the top of the *mammillatum* Zone. Using this definition both the Woburn Sands and the Carstone of the Ely district fall within the Lower Greensand.

There has been much misunderstanding in the past concerning the stratigraphical relationship of the Woburn Sands to the Carstone. Both formations have been referred to as 'Lower Greensand' on earlier maps of the district and the implication that they were contemporaneous has led to confusion. The two formations are lithologically distinct and appear from the published geological maps to occupy mutually exclusive geographical areas, the junction between these areas falling in the middle of the present district at about the Little Ouse River. Attempts have been made to explain this present-day distribution as due to a lateral facies change of the Woburn Sands into the Carstone (e.g. Chatwin, 1948) or to tectonic structure (e.g. the 'Southery Hinge' of Kelly, unpublished PhD thesis, London 1977).

The discovery of Aptian ammonites in the basal bed of the Woburn Sands at Upware and in the Carstone at Hunstanton led several authors, including Lamplugh (*in* Whitaker and Jukes-Browne, 1899) and Spath (1924) to believe that the two deposits were the same age. Keeping (1883) had already suggested that both faunas were derived from older strata, and this was largely confirmed by Casey (1961b, pp.569–571). Because of the unfossiliferous nature of the Woburn Sands and Carstone little advance was made in determining their stratigraphical relationship until relatively recently.

The Woburn Sands rest unconformably on Jurassic or older strata throughout the present district and the adjacent areas. The Carstone in turn rests unconformably on the Woburn Sands and, in the northern half of the Ely district, oversteps them and comes to rest on the Sandringham Sands (Figure 19). To the north of the present district, beds that were deposited contemporaneously with the Woburn Sands reappear from beneath the basal unconformity of the Carstone in north-west Norfolk and Lincolnshire as the Sutterby Marl, a soft calcareous marine mudstone (Figure 20A). At its maximum development, in the Hunstanton district, the Carstone contains ferruginous ooids and many pebbles derived from Lower Cretaceous and older rocks that occur in north Norfolk and Lincolnshire. Both the ooids and pebbles appear to have spread southwards into the Ely district (Figure 20B) which lay close to the southern limit of a

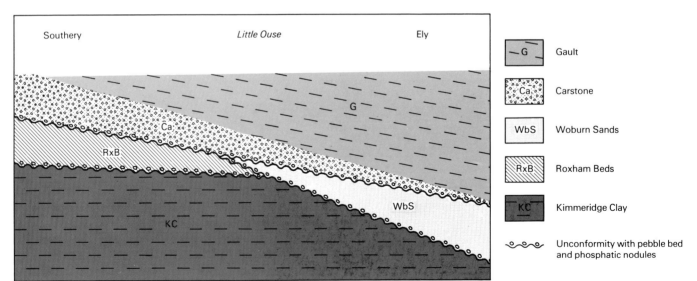

NOT TO SCALE

Figure 19   Relationships of late Jurassic–Cretaceous unconformities in the district

depositional basin that stretched northwards to Yorkshire and beyond.

It is, therefore, likely that the present distributions of the Woburn Sands and Carstone owe much to their depositional distribution and do not result from any contemporaneous or subsequent tectonic event. The area of overlap of the two formations is much larger than that indicated on the published maps because, in the southern part of the Ely district and over much of the Cambridge district, a thin development of Carstone has been included with the Woburn Sands. Thicker developments of Carstone are locally present in Oxfordshire (Gallois and Worssam, 1983), Wiltshire and the Isle of Wight.

WOBURN SANDS

The main outcrop of the Woburn Sands in the district runs from the Little Ouse River to Soham, but is almost entirely obscured by the Recent deposits of the South Level. A large outlier of the formation gives rise to the well-drained sandy plateau on which Ely Cathedral and much of the town is built: smaller outliers form local eminences near Witchford and at Stuntney. Up to 4.6 m of Woburn Sands have been proved in boreholes at Ely, where the maximum total thickness is probably about 5 m. A similar thickness is probably present at Eye Hill Farm [575 767], at the southern end of the main outcrop, but from there the formation thins steadily northwards.

The formation is composed of medium- and coarse-grained glauconitic greenish grey sands and pebbly sands that weather to grey, greenish yellow and rusty yellowish brown. They are locally in part calcareously, ferruginously or siliceously cemented, and in the past these harder beds have been used on a small scale as building stone. Where unweathered the sands are commonly cross-bedded. The formation rests unconformably and with marked lithological contrast on the Kimmeridge Clay everywhere south of the Little Ouse, the junction being marked by a pebble bed containing phosphatised fossils derived from late Jurassic and early Cretaceous strata. At Temple Farm [625 870], Little Ouse, the Woburn Sands overlap onto the Roxham Beds before they are themselves cut out by the overlying Carstone (Figure 19).

The Woburn Sands are sparsely fossiliferous and have yielded only a few ammonites, belemnites and bivalves; wood fragments are locally common. The basal pebble bed contains a great variety of derived fossils (see Casey *in* Worssam and Taylor, 1969, pp.26 – 27 for list). At Upware, Keeping (1883) described a fauna rich in ammonites, bivalves, brachiopods, bryozoans, and gastropods from pebbles and nodules in the basal bed of the Woburn Sands. Much of the fauna was preserved in dense phosphate and was clearly derived from the Oxford, Ampthill and Kimmeridge clays; other parts of the fauna appear to have been derived from the Sandringham Sands and Dersingham Beds of Norfolk.

A large part of the fauna, however, was preserved in softer phosphate and was thought to be indigenous. Casey (*in*

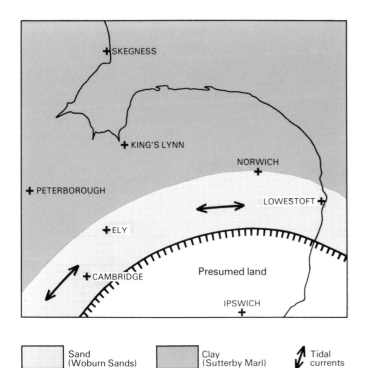

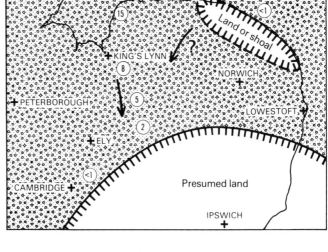

A. Mid-Aptian (*nutfieldiensis* Zone)

B. Lower Albian (*mammillatum* Zone)

Figure 20 Presumed palaeogeography at the time of deposition of the Woburn Sands and Carstone

Worssam and Taylor, 1969) described this part of the fauna as wholly or partly reworked, its phosphatised ammonites being representative of the *Prodeshayesites fissicostatus*, *Deshayesites forbesi*, *D. deshayesi* and *Tropaeum bowerbanki* zones of the Lower Aptian.

Two indigenous ammonites were present, both preserved as internal moulds made of pebbly calcareous sandstone with some small patches of the original shell attached. These were determined by Casey (1961b, p.569) as *Colombiceras sp. nov. cf. tobleri* (Jacob) (Upper Aptian, probably *nutfieldiensis* Zone) and *Tropaeum keepingi* Casey (?Lower Aptian). Unfortunately, neither species has yet been recorded elsewhere in Britain so their precise stratigraphical significance cannot be determined. The recent discovery of a specimen of *Dufrenoyia* in a similar preservation in the Woburn Sands at Haddenham (Dr C.L. Forbes, personal communication) also suggests a late Lower Aptian or early Upper Aptian age. The only other indigenous fossils of stratigraphical significance from the Woburn Sands at Upware are specimens of the Aptian belemnite *Neohibolites ewaldi* (von Strombeck), a species that is common in the Sutterby Marl of Lincolnshire where it occurs with a few indigenous ammonites, including *Colombiceras* and *Dufrenoyia*, indicative of a late Lower Aptian or early Upper Aptian age. It is, therefore, clear that the Woburn Sands in the Cambridge district, and probably throughout the Ely district, are Aptian in age and it seems probable that they belong with either the *Cheloniceras martinioides* Zone or the *nutfieldiensis* Zone.

The sedimentological evidence for the source of the Woburn Sands is inconclusive. The cross-bedding at outcrops in the Buckinghamshire to the Cambridgeshire area have been studied by Schwartzacher (1953) and Narayan (1963) who respectively concluded that the sand had been transported from the north and the north-west. Narayan (1971) interpreted the cross-bedding as that of sandwaves moving parallel to a shoreline that trended NW–SE with land on its south-western side. This conclusion is at variance with the evidence of the lithological nature and distribution of the Woburn Sands, which appear to be shallow-water marine and beach deposits banked up against a shoreline of Palaeozoic rocks (the western edge of the London Platform) that trended NE–SW in Bedfordshire and Cambridgeshire.

Anderton and others (1979, fig. 15.7) have suggested that the Woburn Sands were deposited in a strait between the London Platform and a land area that covered much of the east Midlands and north-east England. This interpretation ignores the presence of the Sutterby Marl, over much of this supposed land area.

The problem of interpretation has arisen largely because the nature and mode of movement of the sand bodies that make up the Woburn Sands has yet to be clearly demonstrated. They may have been sand-waves aligned at high angles to the coastline, as described by Stride (1963) from parts of the English Channel at the present time, or sand-ridges aligned roughly parallel to the coastline, such as those in parts of the southern North Sea (Houbolt, 1968). Until such time as the sedimentological evidence is better understood, the suggestion by Middlemiss (1967, 1976) that the Woburn Sands shoreline lay just south of a line from Brickhill, Buckinghamshire to Upware, Cambridgeshire, seems the most satisfactory in terms of the lithological and

faunal evidence. Assuming this to be correct the palaeogeography in mid-Aptian times can be summarised as shown in Figure 20A.

Rastall (1919, pp.216–217), in a study of the heavy minerals of the Lower Cretaceous of eastern England, found that a calcareous sandstone from the Woburn Sands at Roslyn Hole [554 807], Ely, consisted of roughly equal proportions of quartz sand and tiny pebbles of cherty and ferruginous rocks. The heavy minerals consisted of abundant zircon, common tourmaline and kyanite, staurolite, small amounts of rutile, and rare garnet and blue amphibole. All the commoner species were present as large fresh grains as well as small worn fragments. Garnet and blue amphibole were noted to be abundant in the Sandringham Sands, rare in the Woburn Sands at Ely, and very rare in more southerly outcrops of the Woburn Sands. It seems likely, therefore, that these two minerals are secondarily derived from the Sandringham Sands. Among the more readily identifiable pebbles and derived grains in the Woburn Sands are limonitised clay ironstones and ooliths, phosphates, and glauconites, probably all from the Sandringham Sands.

Sphene, tourmaline and possible cassiterite were recorded by Rastall as more common in the more south-westerly (Bedfordshire) outcrops of the Woburn Sands, and he suggested that these minerals might indicate a source area in south-west England. However, large fresh angular grains of kyanite were common in all samples of Woburn Sands, and Rastall suggested these might have had a Scandinavian origin. The Woburn Sands weather at outcrop to a loamy ferruginous soil in which Seale (1956) found kyanite and staurolite to be prominent constituents of the heavy-mineral assemblage.

Heavy-mineral assemblages of broadly similar composition to those in the Woburn Sands occur in the Sandringham Sands and Carstone (Rastall, 1919) and in the Spilsby Sandstone of Lincolnshire (Ingham, 1929). This suggests that all these formations have a common distant source (or sources). The possibility that the younger strata have obtained their heavy minerals from the older by erosion can be largely discounted because of the size and freshness of many of the mineral grains at all stratigraphical levels.

In a subsequent study Rastall (1925) concluded that the petrography of the Woburn Sands suggested derivation from Millstone Grit or Coal Measures rather than from Lower Palaeozoic or Precambrian rocks. Boswell (1927) considered the pink and deep purple zircons that had been noted by Rastall in the Woburn Sands, to be characteristic of these sands. He believed the kyanites were too fresh to have travelled far and suggested a nearby, but as yet unknown, source.

Sources as far distant as Scandinavia can be discounted now that the presence of continous thick sequences of predominantly argillaceous Lower Cretaceous deposits are known from the intervening North Sea. The likelihood is, therefore, that these heavy minerals were derived from igneous and metamorphic source rocks not far distant from the Woburn Sands outcrop, as suggested by Boswell (1927). The simplest solution, that these rocks now lie buried beneath the Cretaceous rocks of the London Platform, seems unlikely because every borehole drilled there to date has proved Palaeozoic sediments. The recent discoveries of coarse clastic

Carboniferous rocks around the edges of the London Platform in the southern North Sea and in Berkshire and Oxfordshire have added weight to Rastall's (1925) suggestion that much of the quartz in the Woburn Sands could have been derived from this source. The London Platform is also likely to have been fringed by coarse Triassic sediments that were available to be eroded in the early Cretaceous.

DETAILS

Ely

The junction of the Woburn Sands and the Kimmeridge Clay is marked by a prominent feature on the east side of the large outlier at Ely: it is this feature that gives the medieval part of the city its commanding position over the fens of the South Level. Rastall (1919) noted that the Woburn Sands in the vicinity of the cathedral were up to 3 m thick and consisted of pebbly sandstones that had been used in the construction of the cathedral buildings. This type of material can be seen infilling the cathedral walls, notably near the south trancept, and was presumably dug from the immediately adjacent area.

Boreholes at Bray's Lane [5444 8053] and Market Street [5428 8043], Ely, proved 3.9 and 4.6 m of Woburn Sands respectively, resting on Kimmeridge Clay. A sample from 0.9 m above the base of the formation in the first of these boreholes consists of brown, partially weathered, calcite-cemented fine-grained pebbly sandstone with abundant flattened, well rounded pebbles of limonitised clay ironstone up to 10 mm in length, common large angular grains of clear quartz, rare well rounded black chert (lydite), black phosphate and common pitted glauconite grains. The basal bed of the Woburn Sands has been exposed from time to time in the nearby borrow pit at Roslyn Hole [552 807], where it consists of deeply weathered, very pebbly, ferruginous sand that rests on an irregular surface cut into the Kimmeridge Clay. The pebbles include all the types recorded at Bray's Lane and appear to have been largely derived from the Sandringham Sands. Even the phosphatic pebbles, some of which are casts of Kimmeridge Clay fossils, are likely to have come via the Sandringham Sands, as they are more waterworn than those that occur in the basal bed of the Sandringham Sands.

A small outlier of Woburn Sands at Bedwell Hay Farm [520 775], at the southern end of the Ely ridge, is largely obscured by till.

Little Ouse River to Soham

Two boreholes [6292 8685] and [6285 8726] at Temple Farm, Little Ouse proved 3.3 and 4.4 m of Woburn Sands consisting of weathered fine- and medium-grained, pebbly yellow and orange-brown ferruginous sands that passed down into green (glauconite) and brown speckled (tiny ironstone pebbles) grey sands with patchy calcitic cement. The higher beds may include the basal part of the Carstone, but the samples were too deeply weathered for this to be ascertained. A sample from close to the base of the formation in the second borehole, consisted of calcite-cemented, poorly sorted, angular, fine- and medium-grained quartz sand with numerous tiny 'pebbles' (coarse sand grade) of rusty brown limonitised ironstone, limonitised ooids, pitted and well rounded glauconites and well rounded vein quartz (probably all derived from the Sandringham Sands). The rock is permeated by subhorizontal, tube-shaped burrows lined with carbonaceous clay.

Between Temple Farm and St John's Farm [594 778], Soham the outcrop of the Woburn Sands is overlain by Recent deposits. The formation was proved beneath Gault in boreholes at Isleham Fen Pumping Station [6280 7881] and Cooks Drove Farm [6469 7838]

where it provides water supplies. The formation gives rise to sandy ferruginous soils between St John's Farm and Eye Hill Farm [576 766] and on the tops of Stuntney [558 780] and Nornea [570 778] hills where it forms small outliers.

CARSTONE

In the Ely district the Carstone consists of oolitic (chamosite and limonitised chamosite) clayey sand, dark greyish green when fresh, that weathers to a rusty brown sandy clay. The outcrop of the formation can be traced southwards from the cliff exposures at Hunstanton [TF 672 413], Norfolk as far as the Little Ouse River in the present district. Throughout this outcrop the formation rests unconformably on Jurassic and Cretaceous rocks, the age of the underlying beds increasing southwards from Barremian (Roach) at Hunstanton to Kimmeridgian (Lower Kimmeridge Clay) at the Little Ouse. Where unweathered, the Carstone is readily distinguishable from the Sandringham Sands and the Woburn Sands by its abundance of limonitised ooids and the relative scarcity of glauconite and from the Woburn Sands, by its lack of calcareous cement. Where weathered, the limonite and glauconite of all three deposits gives rise to rusty brown and yellow sands that are difficult to distinguish from one another. The outcrop does not form a topographical feature in the district, and it is overlain by Recent deposits everywhere except for two small areas at Stubb's Hill [645 940] where it has been revealed by peat wastage.

The Carstone has not been recognised as a separate formation in previous geological surveys of the district, having been included in the Sandringham Sands in the area north of the Little Ouse and in the Lower Greensand (= Woburn Sands) in the area to the south. Boreholes at Southery and at Soham suggest that the Carstone thins southwards across the district from about 3 m to less than 1 m (Figure 20B). The formation is probably present throughout the Cambridge (Sheet 188) district, but is there everywhere less than 0.5 m thick and has been included with the Woburn Sands.

Fossils other than plant debris are rare in the Carstone and its age was in doubt until relatively recently. At Hunstanton, Casey (1961b, p.571) showed that phosphatised ammonites that occur at the base of the formation were derived from two zones of the Lower Aptian. The first undoubtedly indigenous fossils, other than plant fragments, to be recorded from the Carstone in Norfolk were terebratulid brachiopods collected by Casey (1967, p.91) from near West Dereham (Sheet 159) [TL 662 996] and considered by him to be conspecific with forms from the Lower Albian Shenley Limestone of Bedfordshire. Other brachiopods collected by Casey from the same locality included specimens of *Burrirhynchia leightonensis* (Walker) (Owen and others, 1968, p.518), a species known only from the Shenley Limestone and the Carstone of Lincolnshire. Casey (in Casey and Gallois, 1973, p.11) also recorded the ammonites *Beudanticeras newtoni* Casey, *Douvilleiceras mammillatum* (Schlotheim) and *Leymeriella sp.*, indicative of the *tardefurcata* and *mammillatum* zones of the Lower Albian, from phosphatic nodule beds in the top part of the Carstone at West Dereham. Teall (1875) had recorded a similar fauna from the same nodule beds, but including *Sonneratia kitchini* Spath (indicative of the *mammillatum* Zone), from phosphate workings in the same area. It is difficult to determine how closely these ammonites reflect the age of the deposit because, although they are completely phosphatised, they appear to be little worn even though they are preserved in an abrasive, pebbly sand matrix. It seems reasonable to assume that this part of the Carstone is of *mammillatum* or post-*mammillatum* Zone age. The Gault at West Dereham, and throughout the Ely district, overlies the Carstone without any obvious major break in sedimentation. The presence of abundant *Hoplites spp.*, including *Hoplites (H). dentatus* Spath, in the lowest part of the Gault suggests a *mammillatum* or early *Hoplites dentatus* Zone age for the Carstone there. At Hunstanton, the base of the Carstone can be dated as post-Lower Aptian on the basis of the youngest (*bowerbanki* Zone) derived ammonites in its basal bed. There too, the transitional contact with the overlying Red Chalk, which contains indigenous *Hoplites dentatus* Zone ammonites, suggests that the Carstone is late Lower or early Middle Albian in age.

DETAILS

Southery to the Little Ouse River

A borehole [6490 6463] at Decoy Farm, Southery, drilled in 1975 to examine the supposed alluvium proved by the 1920 oil shale exploration borehole drilled at the same site (Pringle, 1923, p.127), established 2.7 m of Carstone resting on Roxham Beds. The Carstone consists of mottled orange-brown and pale grey soft clayey ferruginous fine-grained sandstone with lenses rich in small pebbles of ironstone and vein quartz. The lowest 0.45 m is uniformly more pebbly and includes, at its base, phosphatic pebbles derived from the underlying Roxham Beds.

A cored site-investigation borehole at Blackdyke Farm, Hockwold [6916 8817] proved about 2.5 m of Carstone overlain by Gault and underlain by calcareous sandstone (Woburn Sands). The Carstone consists of intensely bioturbated, soft, fine-grained sandstone with much disseminated limonite and limonite ooids; burrowfills of phosphatised mudstone in the top part of the bed probably extend down from the overlying Gault. The junction with the Woburn Sands lies at an irregular burrowed surface resting on dense calcite-cemented sandstone. Other boreholes drilled as part of the same investigation between Blackdyke Farm and Holmsey Green [6916 8817 to 6962 7858] penetrated up to 1 m of bioturbated, soft, limonitic sandstone at the top of the formation.

A borehole at Shrubhill Farm [6614 8788] proved 1.7 m of Carstone overlain by Gault. The Carstone consists of pale, slightly greenish grey clay and fine-grained sand with small pebbles of ironstone and vein quartz, and with burrowfills of dark grey clay; a band of sandy and pebbly phosphatic burrowfill nodules occurs close below the junction with the Gault.

Little Ouse River to Soham

No outcrop of Carstone has yet been recorded south of the Little Ouse River, but the presence of specimens of its characteristic lithology at 22.45 to 23.09 m in the Soham Borehole, and of sand with abundant limonite ooids and iron-stained vein quartz pebbles (both typical of the Carstone) immediately beneath the Gault in other boreholes in Cambridgeshire, suggest that a thin (less than 1m) development of Carstone is present everywhere in the Ely district as well as beneath much of the Cambridge district. A record of 'Lower Greensand' in the floor of a borrow pit in the Gault at Castles Farm, near Soham [600 773] was probably Carstone. RWG

GAULT

The Gault was first mapped out as a discrete formation in the Ely district by William Smith who referred to it as the 'Golt Brick Earth' on his geological map of England and Wales (1815) and was subsequently shown in more detail on his geological maps of the counties (e.g. that of Norfolk, 1819). Parts of the formation were exposed in pits dug for brickmaking and agricultural phosphate in Cambridgeshire and Suffolk in the 19th century, and in the present district it has been worked from time to time for embanking materials.

The Gault has a broad outcrop in the Ely district, running from Methwold Fens southwards to Isleham Fen but, with the exception of small areas at Stubbs Hill [645 936], Shrubhill Farm [662 880], Little Ouse [640 868] and the southern part of the district around Broad Hill [592 767], this outcrop is entirely obscured by Quaternary deposits. Where exposed, the Gault weathers to a heavy yellowish brown clay.

The stratigraphy of the formation in the district is known in detail from continuously cored boreholes drilled for oil shale exploration in Methwold Fens and for site investigation in the Hockwold to Mildenhall area. The base of the formation is marked by a rapid, but apparently conformable, lithological transition from the pebbly sands of the Carstone to soft mudstones. The sequence is attenuated in comparison with areas elsewhere in southern England due to its proximity to the London Platform, which lay to the south-east, and the shallows of the Red Chalk sea in the north. The formation is richly fossiliferous, ammonites and bivalves being especially common, and appears to have been deposited in a warm, relatively shallow embayment of the sea. The Gault becomes progressively more calcareous upwards: the highest beds are rich in calcareous microfossils, including coccoliths, and form a lithological transition to the overlying Chalk. Nowhere in the present district, however, does the Gault become sufficiently calcareous or lithified to form even a soft limestone, and is thus readily distinguishable from the Chalk.

The Gault can be divided into two parts on its gross lithology; these divisions were termed the Lower and Upper Gault by De Rance (1868). The Lower Gault consists predominantly of medium and dark grey soft mudstones and silty mudstones in which, according to analyses quoted by Perrin (1971, pp.140–149), illite and kaolinite are the dominant clay minerals. Samuels (1975, p.246) and Jeans (1978, fig. 4) obtained similar results for samples of Lower Gault from a borehole at Blackdyke Farm [6916 8817], Hockwold. The Upper Gault consists largely of pale grey mudstones with high calcium carbonate contents, in which smectite is the dominant clay mineral (Perrin, 1971).

The Lower Gault and the lower part of the Upper Gault of the Ely district are made up of a number of small-scale rhythms, mostly 1 to 2 m thick in the south-west Norfolk sequence described below. Each rhythm begins with a medium or dark grey, gritty (shell debris, mostly *Inoceramus* prisms), shelly (abundant oysters and common belemnites), pebbly (exhumed phosphatised burrowfills, rolled phosphatised ammonites and bivalves) silty mudstone or muddy siltstone that rests on a partially phosphatised and glauconitised, burrowed surface. These silty basal beds pass up into medium and pale grey calcareous mudstones by a decrease in coarser clastic (including bioclastic) material and an increase in calcium carbonate (Figure 21). This lithological change is commonly accompanied by a decrease in faunal diversity and numbers. The middle and upper parts of the Upper Gault show weak rhythms, are uniformly paler and more calcareous than the Lower Gault, and have fewer major erosion surfaces.

The Gault appears to thin steadily northwards across the Ely district, from about 25 m in the Soham area to about 18 m beneath Methwold Fens. This thinning is related to a corresponding thickening in the underlying Carstone, and it has been suggested that the Gault of the Ely district infilled a broad submarine hollow formed between the London Platform land area in the south and an off-shore bar composed of Carstone in the north (Gallois and Morter, 1982).

The Gault contains a rich marine fauna (Plate 5) dominated by bivalves but including relatively common solitary corals, serpulids, gastropods, scaphopods, ammonites, belemnites, crinoids and echinoids, together with an abundant microfauna of ostracods and foraminifera. Ichnofossils are represented by common *Chondrites*, possible thalassinoid burrows and a variety of trails. Extensive collections, especially of ammonites and bivalves, were made from the numerous, generally small, exposures available to collectors in Victorian times. These collections are commonly poorly documented stratigraphically, because the exposures were separated from one another by long stretches where the mudstones are deeply weathered and rarely, if ever, exposed. Most stratigraphical studies on the Gault have been concentrated on the cliff and foreshore exposures of the type section at and adjacent to Folkestone, Kent. There, the whole of the formation has been available for study from time to time, albeit in sections separated by landslip. De Rance (1868, *in* Topley 1875), Price (1874, 1875, 1879), Jukes-Browne (1900), Spath (1923–43), Casey (1961b) and Owen (1971, 1975) all worked on these sections, and the succession of ammonite assemblages on which the formation is zoned is now known there in considerable detail.

Little has been written on the Gault at outcrop in Cambridgeshire and Norfolk since Jukes-Browne's review in the Cretaceous Rocks of Britain (Vol. 1, 1900) in which the local details are based largely on those given in the various sheet memoirs for the area. Most of the Cambridgeshire Gault was believed to belong with the Lower Gault (Jukes-Browne, 1900, p.287). A number of critical sections in Norfolk, where the succession was poorly exposed but was believed to be more complete than in Cambridgeshire, were described by Jukes-Browne and Hill (1887). Descriptions of several small sections in the Gault in the Huntingdon, Biggleswade and Cambridge districts are given in Geological Survey memoirs (Edmonds and Dinham, 1965; Worssam and Taylor, 1969).

Attempts have been made to correlate the Gault sequence in East Anglia with that at Folkestone but because of the scattered nature of the East Anglian exposures these attempts have, until recently, mostly been limited to the correlation of individual sections with part of the Folkestone sequence. In recent years, however, continuously cored boreholes have been drilled though all or part of the Gault in the Ely and adjacent districts and it has become apparent that lithological and faunal marker bands can be traced throughout East Anglia and parts of southern England. The main obstacle to making correlations within the Gault is the difficulty in recognising the bases of those individual rhythms that are erosive, and thus locally cut out all or part of the subjacent rhythms. Work on Gault sequences in continuously cored boreholes has shown that correlations can be made with confidence over relatively large distances in East Anglia by comparing the sequence of lithological and faunal events locality by locality. These comparisons allow recognition of a standard succession of 19 distinctive beds (G1 to 19) in the Gault of the central and northern parts of East Anglia (Gallois and Morter, 1982). All but the highest of these beds have been recognised in the Ely district. The lithological succession, zonal and subzonal scheme, together with selected marker bands and faunal ranges are shown in Figure 22. Descriptions of the individual beds, based on the cored boreholes drilled through the Gault of the Hockwold to Mildenhall area are given below.   RWG, AAM

DETAILS

Methwold Fens to the Little Ouse River

The full thickness of the Gault was penetrated in 23 site-investigation boreholes drilled in 1967 between Blackdyke Farm [6916 8817] and Kentford, Suffolk [7024 6838], to explore the route for a tunnel aqueduct in the Gault for the Ely-Ouse Water Transfer Scheme (Samuels, 1975). The eleven most northerly boreholes, Nos. 13 to 23, fall within the present district and have been used to compile the descriptions of the Gault given below. Borehole 12 is within 100 m of the eastern boundary of the district (see Appendix 1 for details). All twelve boreholes were continuously cored although core losses, the removal of samples for engineering testing, and frost damage to some cores has meant that no single core shows the complete sequence. The formation has been divided into eighteen distinctive

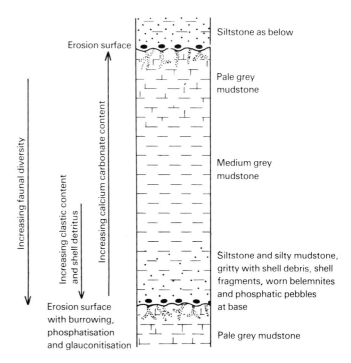

Figure 21 Idealised rhythm for the Gault of the district

beds on the basis of a combination of lithological and faunal characters. Each bed appears to be present throughout the Ely district with the exception of Bed G 6 which is locally removed by erosion at the base of Bed G 7.

The following bed descriptions of the complete sequences between the Carstone and the Cambridge Greensand include a few ammonites not recorded in the Ely district, but which occur in the equivalent beds in a continuously cored BGS borehole at Mundford [Borehole C: 7787 9320] in the Thetford district (Sheet 174), some 10 km north-east of Ely-Ouse Borehole 23. The thicknesses quoted are for Borehole 14 [6962 8115], though the lithologies and faunas summarise the provings in the twelve boreholes. The thickness of Bed G 6, absent in Borehole 14, is that proved in the Mundford C Borehole.

Cambridge Greensand: see p.62 for details

Stoliczkaia dispar Zone:

BED G 18: mudstone; very pale grey, smooth textured with irregular fracture; bioturbated throughout with *Chondrites* at several levels picked out by slightly silty mudstone infillings; pale brown, soft phosphatic burrowfills common; sparsely shelly with *Aucellina spp.* including *A.* gr. *gryphaeoides* (J. de C. Sowerby) and *A. uerpmanni* Polutoff group, *Entolium orbiculare* (J. Sowerby) and *Plicatula gurgitis* Pictet & Roux the only common bivalves; *Plagiostoma globosa* (J. de C. Sowerby), *Pycnodonte (Costeina) sp.* and *Holaster* cf. *laevis* de Luc also common; small belemnites, including *Neohibolites praeultimus* Spaeth, relatively common; rare ammonites include *Anisoceras spp.*, *Cantabrigites cantabrigense* (Spath); *C. minor* (Spath), *Callihoplites glossonotus* (Seeley), *Lechites sp.*, *Lepthoplites* cf. *pseudoplanus* Spath and *Mortoniceras (M.) sp.*, pyritised burrows and trails common; fish debris common; base taken at change to *Aucellina*-rich mudstones   1.9 m

BED G 17: mudstone; very pale grey, smooth textured, friable; shelly with abundant *Aucellina* spp. of *gryphaeoides* and *uerpmanni* groups; bioturbated throughout; sparse other fauna includes *Entolium orbiculare* and other thin-shelled bivalves; *Anisoceras* cf. *exoticum* Spath, *Callihoplites* including *C.* cf. *cratus* (Seeley), *C. glossonotus*, *C.* cf. *leptus* (Seeley) and *C.* cf. *pulcher* Spath, *Lepthoplites spp.* including *L. cantabrigiensis* Spath, and *Neohibolites* including *N. praeultimus* and rare *N. minimus* (Miller), also present; pyritised (pins) burrowfills common; base taken at change to sparsely fossiliferous mudstones and commonly marked by a burrowed surface with phosphatisation and black phosphatic pebbles   0.9 m

Mortoniceras inflatum Zone

BED G 16: mudstone; pale grey becoming darker with depth, smooth textured; almost barren except for scattered *Aucellina* in the upper part and rare *Neohibolites minimus*; shelly pebble bed 2 to 3 cm thick at base (the Milton Brachiopod Band) consisting of phosphatic burrowfills and phosphatic pebbles (mostly 5 to 10 mm across) with thick cream-coloured cortices and hard dark brown centres, set in a matrix of silty and gritty (shell debris) medium grey mudstone; pebble bed fauna includes common *Moutonithyris dutempleana* (d'Orbigny) and *Terebratulina* cf. *martiniana* (d'Orbigny), common oysters including *Pycnodonte (Costeina) sp.*, *Mortoniceras (M). sp.* and rare phosphatised ammonites; irregular burrowed surface at base   2.1 m

BED G 15: mudstone; pale grey, smooth textured in upper part becoming slightly silty and silty with depth; bioturbated in upper part with *Chondrites* and other burrows picked out in darker mudstone; moderately shelly with '*Inoceramus*' *lissa* (Seeley) common throughout; rich ammonite fauna including *Callihoplites* cf. *pulcher*, *C.* cf. *strigosus* Spath, *C.* cf. *variabilis* Spath, *Hysteroceras bucklandi* Sparth *Lepthoplites* cf. *falcoides* Spath, '*L.*' cf. *ornatus* Spath, common large *Mortoniceras (M)* including *M. M.) fissicostatum* Spath and *M. M.) inflatum* (J. Sowerby), *Prohysteroceras (Goodhallites)* and *Stomohamites* cf. *subvirgulatus* Spath; brachiopods including *Kingena*

spinulosa (Davidson & Morris) and *Moutonithyris dutempleana* also present; sparse bivalve fauna includes rare *Aucellina coquandiana* (d'Orbigny), *Callicymbula* cf. *phaseolina* (Michelin), *Entolium orbiculare*, *Plagiostoma globosa*, *Plicatula radiola gurgitis* and relatively common *Pycnodonte (Phygrea) sp.*; rare *Neohibolites* including *N. minimus* and *N. praeultimus* and with *N. ernsti* Spaeth in basal bed; cirripede valves locally common; phosphatised thallasinoid burrowfill in lower part of bed; silty glauconitic bed at base (locally cemented to form a limestone, the Barnwell 'Hard Band') with much gritty shell debris composed of *Inoceramus* prisms and other shells and with abundant ostracods; phosphatised burrowfills and phosphatised *Dentalium* rest on a burrowed surface   1.3 m

BED G 14: mudstone; pale grey, smooth textured with hackly fracture; sparsely shelly but with common *Neohibolites*, including *N. praeultimus* in the upper part, *N. ernsti* and *N. minimus* throughout

Plate 5   Fossils from the Gault of the region

1 *Eopecten studeri* (Pictet & Roux), × 1. Upper Gault *varicosum* Subzone, Burwell Brickworks pit, Cambridgeshire (SCH 1818)
2 *Nucula (Pectinucula) pectinata* J. Sowerby, × 1. Upper Gault *orbignyi* Subzone, Mundford C Bh. (BDN 3685)
3 *Aucellina cycloides* (Polutoff), × 2. Upper Gault *rostratum* Subzone, Mundford C Bh. (BDN 3471)
4 *Aucellina* sp., × 1. Upper Gault, Severals House Bh. (WM 4450)
5 *Hysteroceras* cf. *orbignyi* Spath, × 1. Upper Gault *orbignyi* Subzone, Ely-Ouse Bh. 11 (BPB 4940)
6 *Hoplites (H.)* aff. *spathi* Breistroffer, × 1. Lower Gault, Castles Farm Pit, Soham (FD 2172)
7 *Hysteroceras bucklandi* Spath, × 1. Upper Gault *auritus* Subzone, Burwell Brickworks pit, Cambridgeshire (SCH 2194)
8 '*Inoceramus*' *lissa* (Seeley), × 1. Upper Gault *auritus* Subzone, Duxford Bh. (Bt 588)
9 *Birostrina concentrica* (Parkinson), × 1. Lower Gault *spathi* Subzone, Ely-Ouse Bh. 18 (BDM 9909)
10 *Neohibolites ernsti* Spaeth, × 2. Upper Gault, Milton Borrowpit, Cambridgeshire (GSM 117332)
11 *Neohibolites oxycaudatus* Spaeth, × 2. Upper Gault *orbignyi* Subzone, Duxford Bh. (Bt 668)
12 *Birostrina* cf. *concentrica* subsp. D of Kaufmann, × 1. Upper Gault *varicosum* Subzone, Gayton Bh. (BDB 8488)
13 *Tetraserpula* sp., × 2. Upper Gault *orbignyi* Subzone, Ely-Ouse Bh. 11 (BDP 4969)
14 *Pentaditrupa* sp., × 2. Upper Gault *auritus* Subzone, Ely-Ouse Bh. 2 (BDP 2552)
15 *Anahoplites mantelli* Spath, × 1. Lower Gault *intermedius* Subzone, Mundford C Bh. (BDN 3972)
16, 17 and 18 *Anomia* cf. *carregozica* Maury, × 1. Lower Gault 'Anomia Bed', Ely-Ouse Bh. 9 (BDP 4436, 4441 and 4447)
19 *Birostrina sulcata* (Parkinson), × 1. Upper Gault *orbignyi* Subzone, Ely-Ouse Bh. 11 (BDP 4930)
20 *Hoplites (H.)* aff. *spathi* Breistroffer, × 1. Lower Gault *spathi* Subzone, Castles Farm pit, Soham (Zn 2759)
21 and 26 '*Ostrea*' *papyracea* Sinzow, × 1. Lower Gault ?*lyelli* Subzone, West Dereham, Norfolk (GSM 117279 and 117278)
22 *Cantabrigites minor* (Spath), × 1. Upper Gault, Ely-Ouse Bh. 9 (BDP 4018)
23 *Cantabrigites cantabrigense* (Spath), × 1. Upper Gault, *rostratum* Subzone, Ely-Ouse Bh. 14 (BDP 6166)
24 *Hysteroceras varicosum* (J. Sowerby) *binodosa* Stieller, × 1. Upper Gault *varicosum* Subzone, Ely-Ouse Bh. 18 (BDM 9848)
25 *Nielsenicrinus cretaceus* (Leymerie), × 1. Upper Gault *orbignyi* Subzone, Gayton Bh. (BDB 8526)

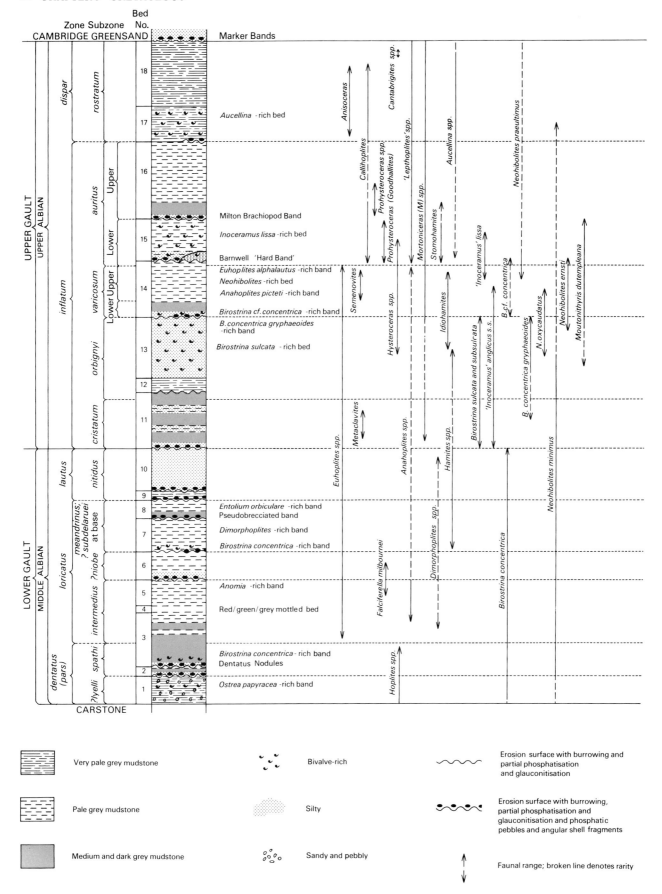

Figure 22 Generalised vertical section of the Gault of the region. Not to scale; see text for thicknesses

and with *N. oxycaudatus* Spaeth in the lower part; relatively common *Euhoplites* including *E. alphalautus* Spath common in top part of bed and *E. vulgaris* Spath; *Hysteroceras binum* (J. Sowerby), *H.* cf. *orbignyi* Spath, *H. varicosum binodosa* Stieler, *Idiohamites* cf. *spinulosus* (J. Sowerby) *I.* cf. *subspiniger* Spath, *Mortoniceras (M) sp.* and *Semenovites sp.* also present; '*Inoceramus*' *lissa* in top part of bed; rare '*Inoceramus*' *anglicus* Woods in lower part and *I.* aff. *anglicus* throughout; *Cyclocyathus fittoni* Milne-Edwards & Haime, *Parsimonia antiquata* (J. de C. Sowerby), *Kingena spinulosa*, *Moutonithyris dutempleana*, *Terebratulina* cf. *martiniana*, *Barbatia marullensis* (d'Orbigny), *Eopecten studeri* Pictet and Roux, *Nucula (Pectinucula) pectinata* J. Sowerby, *Pycnodonte (Phygrea)* aff. *vesicularis* (Lamarck), *Turnus sp.*, cirripede valves, *Nielsenicrinus cretaceus* (Leymerie) and *Stereocidaris gaultina* (Woodward) also present; becoming darker, and more silty in lowest part with common *Chondrites*; very common *Birostrina* cf. *concentrica* Parkinson in basal part; erosion surface, phosphatisation and burrowing at base   1.6 m

BED G 13: mudstone; medium grey with bioturbation, including *Chondrites*, picked out by paler grey burrowfills; slightly silty to silty throughout; very fossiliferous, especially in upper part, with abundant *Birostrina sulcata* (Parkinson) throughout and *B. concentrica gryphaeoides* (J. de C. Sowerby) common in top part of bed; abundant ammonites, mostly *Euhoplites* including *E. armatus* Spath, *E. inornatus* Spath, *E. proboscideus* (J. Sowerby), *E. subcrenatus* Spath and *E. trapezoidalis* Spath, and with *Anahoplites sp.*, *Dipoloceras sp.*, *Hysteroceras binum*, *H. carinatum* Spath, *H. orbignyi*, *Mortoniceras (Deiradoceras) sp.* and hamitids, with *Idiohamites* in upper part and *Hamites* including *H.* cf. *intermedius* J. Sowerby in lower part; *Trochocyathus sp.*, *Parsimonia antiquata*, *Kingena spinulosa*, *Moutonithyris dutempleana*, *Atreta sp.*, *Nucula (Pectinucula) pectinata*, *Plicatula sp.*, *Pycnodonte (Phygrea)* aff. *vesicularis*, *Neohibolites oxycaudatus*, *N. ernsti*, *Nielsenicrinus cretaceus*, *Cirsocerithium subspinosum* (Deshayes) and *Dentalium sp.* also present; '*Inoceramus*' *anglicus* locally common; less fossiliferous in lower part with *Birostrina sulcata* and *B. subsulcata* Wiltshire locally common and with relatively rare ammonites and abundant *Neohibolites minimus* (Miller); burrowed surface at base with phosphatised burrowfills   1.9 m

BED G 12: mudstone; very pale grey with burrowfills, including *Chondrites*, of dark grey mudstone; smooth textured; very calcareous becoming locally weakly cemented; sparsely fossiliferous with *Birostrina sulcata* and '*Inoceramus*' *anglicus* the only common fossils; - *Jurassiphorus fittoni* (Roemer), *Nucula (Pectinucula) pectinata* (J. Sowerby), *Plicatula sp.*, *Pycnodonte (Phygrea)* aff. *vesicularis*, *Euhoplites inornatus*, *E. sp.* (*ochetonotus-sublautus* gr.), *Neohibolites minimus* and *Nielsenicrinus cretaceus* also present; large phosphatised burrowfills at base   0.4 m

BED G 11: mudstone; medium grey with thin pale grey interbeds; slightly silty; bioturbated throughout with *Chondrites* at several levels; moderately fossiliferous with *Birostrina sulcata* locally common, and less common *B. subsulcata*; common *Neohibolites minimus* in translucent preservation; '*Inoceramus*' *anglicus* common; *Birostrina concentrica gryphaeoides* in top part of bed; *Cyclocyathus fittoni*, *Anchura carinata* (Mantell), *Jurassiphorus fittoni*, *Nucula (Pectinucula) pectinata*, *Turnus sp.*, cirripede valves and *Nielsenicrinus cretaceus* also present; the serpulid *Tetraserpula*, *Dentalium (Fissidentalium) decussatum* J. Sowerby and terebelloid burrowfills lined with fish debris, and *Cirsocerithium* locally common; phosphatised burrowfills present at several levels; rare ammonites include *Anahoplites sp.*, *A.* aff. *planus* (Mantell), *Euhoplites* cf. *inornatus* in topmost part of bed, *E. ochetonotus* (Seeley) and *E. trapezoidalis*, *Metaclavites sp.* and *Mortoniceras (M.) sp.*; '*Inoceramus*' *anglicus* present in lowest part; pebble bed at base with rounded and angular hard phosphatic pebbles, some enclosing ammonites, set in silty, shelly mudstone; burrowed junction with   1.5 m

Euhoplites lautus Zone

BED G 10: mudstone; medium grey, slightly silty and silty, moderately shelly with common *Birostrina concentrica* Parkinson and *Neohibolites minimus minimus* Miller in translucent preservation; *Cyclocyathus fittoni*, *Lingula sp.*, *Nucula (Pectinucula) pectinata*, *Anahoplites sp.*, *Dimorphoplites sp.*, *Euhoplites*, including *E. nitidus* Spath and *E.* cf. *opalinus* Spath, *Hamites maximus* J. Sowerby and fish debris also present; phosphatic pebble bed resting on burrowed surface and sharp colour change at base   1.1 m

BED G 9: mudstone; pale and very pale grey and brownish grey; bioturbated throughout with *Chondrites* and other burrow-mottling picked out by darker mudstone infillings; sparsely shelly but with diverse fauna including *Dentalium (Fissidentalium) decussatum* locally common, *Birostrina concentrica*, *Callicymbula phaseolina*, '*Inoceramus*' aff. *anglicus*, *Pinna sp.*, *Neohibolites minimus*, *Dimorphoplites sp.*, *Euhoplites spp.* and *Stereocidaris gaultina*: phosphatic pebble bed at base resting on burrowed surface   0.3 m

Euhoplites loricatus Zone

BED G 8: mudstone; pale and medium grey interbedded and interburrowed; fossiliferous with common *Birostrina concentrica*, *Entolium orbiculare* locally common and *Euhoplites* including *E.* cf. *bilobus* Spath and *E.* cf. *cantianus* Spath; *Cyclocyathus fittoni*, *Callicymbula phaseolina*, *Nucula (Pectinucula) pectinata*, *Turnus sp.*, *Neohibolites minimus minimus*, and fish remains also present; distinctive pebble bed at base composed largely of shell debris (mostly bivalves and belemnites) and numerous small angular pebbles of pale green and, less commonly, reddish brown, soft mudstone; burrowed junction with bed below   0.5 m

BED G 7: mudstone; pale grey, smooth textured with some *Chondrites* and other bioturbation picked out by dark grey mudstone and pale green mudstone; moderately shelly in upper part becoming very shelly in lower part with abundant *Birostrina concentrica*, *Neohibolites minimus minimus*, *Nucula (Pectinucula) pectinata*, *Dimorphoplites* locally very common including *D.* cf. *doris* Spath, *D.* aff. *pinax* Spath and *D. sp. nov.*, and common *Hamites*; '*Trochocyathus*' *conulus* Michelin *non* Phillips, *Kingena spinulosa*, '*Inoceramus*' aff. *anglicus*, *Plicatula sp.* and *Hemiaster* cf. *asterias* Woodward also present; erosion surface at base with *Birostrina concentrica* plasters and phosphatic pebbles resting on burrowed, partially phosphatised surface   1.8 m

BED G 6: mudstone; pale and medium grey and greenish grey becoming darker and silty with depth; moderately shelly with common *Birostrina concentrica*; *Euhoplites loricatus* Spath common in lower part; other fauna includes '*Trochocyathus*' *sp.*, *Neohibolites minimus minimus* and *Falciferella milbournei* Casey; *Birostrina* plaster at base with phosphatic pebbles resting with sharp colour change on burrowed surface   0.9 m

BED G 5: mudstone; pale grey with *Chondrites* and other bioturbation picked out by darker infillings; shelly with *Anomia* cf. *carregozica* Maury abundant in upper part; *Birostrina concentrica* common throughout and with rare *Birostrina concentrica braziliensis* (White); *Neohibolites minimus minimus* and '*attenuatus*' form common in lower part; *Dentalium sp.*, *Entolium orbiculare*, *Anahoplites intermedius* Spath and *Hemiaster sp.* also present; colour change at base   0.5 m

BED G 4: mudstone; strikingly interburrowed pale green, reddish brown and medium grey; sparsely fossiliferous with *Anomia carregozica* and *Birostrina concentrica* the only common fossils; *Bakevellia rostrata* (J. de C. Sowerby) *Birostrina concentrica*, *Anahoplites mantelli* Spath, *Dimorphoplites spp.* and *Neohibolites minimus* also present; colour change at base   0.6 m

BED G 3: mudstone; pale and medium grey, bioturbated becoming medium grey and slightly silty with depth; silty textured throughout due to abundant shell dust and foraminifera; fossiliferous with crushed, rich, shelly fauna including common *Birostrina concentrica* and '*Ostrea*' *papyracea* Sinzow (especially in basal beds); *Anahoplites* cf. *intermedius* Spath, *Dimorphoplites* including *sp. nov.* aff. *tethydis* Spath and *Euhoplites loricatus* in upper part of bed and *Hoplites*, including *H. (H.)* aff. *vectense* Spath, in lower part; *Moutonithyris sp.*, *Anticonulus conoideus* (J. de C. Sowerby), *Rissoina sowerbii* J.S. Gardner, *Entolium orbiculare*, '*Inoceramus*' aff. *anglicus*, *Pycnodonte (Phygrea) sp.*, *Ludbrookia tenuicosta* (J. de C. Sowerby), *Neithea* including *N.* aff. *quinquecostata* (J. Sowerby), *Nucula (Pectinucula) pectinata*, *Neohibolites minimus minimus* and *N. minimus pinguis* Stolley, *Hemiaster baylei* Woodward and other echinoid fragments, also present; pyritised *Birostrina* and *Hoplites* common in lower part of bed; the *dentatus/loricatus* zonal boundary falls in middle part of bed; pebble bed at base set in shelly and gritty, silty mudstone with hard black phosphatic pebbles with common phosphatised hoplitid fragments (the Dentatus Nodules), including *H. (H.)* cf. *dentatus* (J. Sowerby) and *H. (H.)* cf. *spathi* Breistroffer, resting on burrowed surface   1.5 m

Hoplites dentatus Zone (pars)

BED G 2: mudstone, pale and medium grey with bioturbation picked out by paler grey *Chondrites* and other burrowfills; shell dust common throughout; fossiliferous with common *Birostrina concentrica* and common '*Ostrea*' *papyracea*; *Cyclocyathus fittoni*, *Kingena spinulosa*?, *Moutonithyris dutempleana*, *Tamarella* cf. *oweni* Peybernes & Calzada, cf. *Mesosaccella woodsi* (Saveliev), *Nucula (Pectinucula) pectinata*, *Pseudolimea gaultina* Woods, *Pycnodonte (Phygrea) sp.*, *Rastellum sp.*, *Neohibolites minimus* and coarsely ribbed *Hoplites spp.* (*H.) dentatus* and *H. (H.) spathi*, also present; *Birostrina* plaster and phosphatic pebble bed at base set in sandy and silty mudstone and resting on a burrowed surface of   0.2 m

BED G 1: mudstone; pale and very pale, slightly brownish grey; intensely bioturbated with burrowfills of brown sand, limonite ooids and small shiny brown-coated pebbles present throughout; sparsely fossiliferous with *Birostrina concentrica*, *Pycondonte (Phygrea) sp.*, '*Ostrea*' *papyracea* locally very common, *Neohibolites minimus* and coarsely ribbed *Hoplites spp.* including *H. (H.)* cf. *pseudodeluci* Spath; sand and pebble content increasing rapidly with depth; passing down with intense bioturbation into the Carstone   0.8 m

Carstone: see p.50 for details

The full thickness of the Gault in boreholes 12 to 23 ranges from 15 to 20 m (see Appendix 1 for details).

The partially cored boreholes at Methwold Common [678 941] and Several's House [6921 9639] proved Gault beneath Recent deposits and Lower Chalk respectively. The first proved 11.0 m of Gault: the second proved the full thickness of the formation to be 18.2 m (Pringle, 1923, pp.128–129). No specimen has survived from the Methwold Common Borehole, but fragments of fossiliferous mudstone from Several's House include specimens of *Birostrina sulcata* (Parkinson) and *B. concentrica gryphaeoides* (J. de C. Sowerby) indicative of the *orbignyi* Subzone (Bed G 13), *Moutonithyris dutempleana* (d'Orbigny) and *Birostrina* cf. *concentrica* fragments, probably from Bed G 14, and abundant *Aucellina spp.* in typical Bed G 17 lithology.

Very pale calcareous clays in the upper part of the Gault crop out on the low rise and in the ditches adjacent to Shrubhill Farm [663 881]. A percussion borehole drilled near the crest of the rise [6614 8788] in 1975 proved Pleistocene drift on 19.9 m of Gault resting on Carstone. A sample from near the base of the Gault yielded common *Birostrina concentrica* (Parkinson), *Kingena spinulosa* (Davidson & Morris), *Neohibolites minimus* (Miller), and phosphatized fragments of *Hoplites (H.) dentatus densicostata* Spath, *H. (H.)* aff. *persulcatus* Spath and *H. (H.) spp.* : these latter are from the Dentatus Nodules (Bed G 3). A water borehole drilled at the farm [6618 8806] proved 23.4 m of Gault on Carstone. A nearby borehole at Corkway Drove

[6763 8972] proved peat on 22.2 m of Gault. These thicknesses suggest that all three boreholes were sited close to the top of the Gault.

Little Ouse River to Soham

The Gault crops out as a long strip of heavy clay ground on the south side of the Little Ouse River between Temple Farm [632 872] and Decoy Farm [654 863]. The Great Ouse River Authority formerly worked the Gault in a borrow pit near the eastern end of this outcrop [652 865] where the following fauna was collected in 1951: *Cyclocyathus* sp. indet; terebelloid burrowfill with fish debris; *Moutonithyris dutempleana* (d'Orbigny), *Anomia sp. nov.*, *Barbatia marullensis* (d'Orbigny), *Birostrina concentrica*, cf. *Eopecten studieri* (Pictet & Roux), '*Inoceramus*' *anglicus* Woods, *Nucula (Pectinucula) pectinata* (J. Sowerby), *Dentalium (Fissidentalium) decussatum* J. Sowerby, *Euhoplites sublautus* Spath (determined by L. F. Spath, 1951), *E.* aff. *lautus* (J. Sowerby), *E. sp.* indet. (*opalinus* gr. Spath), *Hamites maximus* J. Sowerby, *H. sp.*, hoplitids indet. and common *Neohibolites minimus minimus* (Miller) including '*attenuatus*' forms. This assemblage is representative of the *Dipoloceras cristatum* subzone (Bed G 11); the presence of *Birostrina concentrica* may indicate that the underlying bed (Bed G 10) was also penetrated. Hancock (Sedgwick Museum MS, 1955) subsequently recorded ammonites indicative of the *D. cristatum*, *Hysteroceras orbignyi* and *H. varicosum* subzones (Beds G 11 to G 14) from the same locality. The ammonites include *Beudanticeras sp.* (high zonal form), *Euhoplites alphalautus* Spath, *E. armatus?* gr. Spath, *E.* cf. *boloniensis* Spath, *E. sublautus* Spath, *E. subcrenatus* Spath, *E. trapezoidalis* Spath, *E.* cf. *trapezoidalis*, *E. vulgaris* Spath, *Hamites* aff. *intermedius* J. Sowerby, *Hysteroceras subbinum* Spath, *Idiohamites* cf. *subspiniger* Spath, *Mortoniceras (M.) pricei* Spath, *M. (M.) sp.*, and *Prohysteroceras (Goodhallites)* cf. *goodhalli* J. Sowerby.

Between Decoy Farm and the Soham area, the Gault outcrop is largely overlain by Recent deposits and its position is known only from spoil from the deeper drains and from boreholes at Isleham Fen [6280 7881]. In 1983, a temporary exposure [631 803] in a newly cleaned drain near County Farm proved pale grey Gault clay with phosphatised ammonite fragments beneath about 1.5 m of Quaternary deposits (Dr C. L. Forbes, personal communication). Dr Forbes has identified the ammonites as *Hoplites spp.* including *H. (H.) dentatus* and *H. (H.) rudis* Parona and Bonarelli, and they are clearly indicative of the basal beds of the Gault. The junction with the Lower Greensand was exposed, also beneath drift deposits, in the same drain [at about 629 803], about 600 m E of the position shown for it on the published geological map. Clearly, more information is needed before the position of this boundary can be accurately plotted beneath the South Level. A borehole at Cooks Drove Farm [6469 7838] that reputedly proved 4.3 m of gravel overlying 25.0 m of Gault either encountered a deep, gravel-filled channel that has cut through the Chalk or, more probably, has mistakenly included the sandy and pebbly basal beds of the Chalk (Cambridge Greensand) within the gravel.

Gault was worked from 1947 to 1951 in a borrow pit at Castle's Farm (also known as Crisps' Pit) [600 773], near Soham. Descriptions of the section have been given by Forbes (1960, pp.239–240) and Worssam and Taylor (1969, p.39). The *Hoplites spathi* and *Anahoplites intermedius* subzones (Beds G 1 to G 4) of the Lower Gault were probably exposed and, from time to time, the junction with the Carstone. The lower part of the section contained several horizons of phosphatic nodules and pebbles, one of which was probably the Dentatus Nodules. The following recorded ammonites (determined by L. F. Spath in 1951), mostly preserved in phosphate, indicate sediments or former sediments of *spathi* Subzone age: *Hoplites (H.) dentatus* (J. Sowerby), *H. (H.) dentatus densicostata* Spath, *H. (H.) dentatus robusta* Spath, *H. (H.) paronai* Spath, *H. (H.) persulcatus* Spath, *H. (H.) spathi* Breistroffer and *H. (H.) spp.* The presence of the *intermedius* Subzone is indicated by *Dimorphoplites* aff. *tethydis* (Bayle). Other fauna included *Parsimonia antiquata* (J. de C. Sowerby),

Birostrina cf. *concentrica* (*spathi* Subzone form), *Pycnodonte* (*Phygrea*) aff. *vesicularis* (Lamarck), *Eutrephoceras* cf. *clementinum* (d'Orbigny) and *Neohibolites minimus minimus* (Miller).

The BGS borehole at Soham was sited on Terrace Gravel on Upper Gault (Worssam and Taylor, 1969, p.7). The small number of specimens available for re-examination include the junction with the Carstone at 22.45 m; medium grey and brownish grey mottled mudstone with *Bakevellia rostrata*, *Birostrina concentrica* and *Nucula* (*Pectinucula*) *pectinata* (probably Bed G 4) at 21.41 m; medium grey, silty mudstone with *Euhoplites* cf. *opalinus* and *Birostrina concentrica* (probably Bed G 10) at 17.5 to 17.8 m; silty mudstone with common *Birostrina sulcata*, some partially phosphatised at 16.6 m (basal Bed G 11 and hence the base of Bed G 11 is probably in a core-loss at about 16.7 m); pale and medium grey mudstones with abundant *Hysteroceras*, including *H.* cf. *carinatum* and *H. orbignyi*, *Hamites* cf. *intermedius*, *Euhoplites armatus* and common *E. inornatus*, *Mortoniceras* and abundant *Birostrina sulcata* (Bed G 13) from 12.6 to 13.0 m; the junction of Beds G 13 and G 14 (also the junction of the *orbignyi* and *varicosum* subzones) at 12.26 m with *Birostrina concentrica* common from 11.89 to 12.26 m and with *Euhoplites alphalautus*, *Hysteroceras* cf. *orbignyi*, *?Semenovites sp.*, *Inoceramus anglicus* and *Neilsenicrinus cretaceus* (Leymerie) also present. The beds above 12.26 m up to the highest specimen preserved (at 7.42 m) all appear to lie in Bed G 14. The Gault sequence at Soham is, therefore, much expanded in comparison with that proved in the Ely-Ouse boreholes, the expansion occurring largely within the Upper Gault.

A phosphatised *Hoplites* (*H.*) *persulcatus*, probably derived from the Dentatus Nodules (Bed G 3), was recorded in cavings from the Lakenheath Borehole.

Ely

Skertchly (1877, p.236) recorded Gault overlain by Cambridge Greensand within the erratic mass of Cretaceous rocks formerly exposed at Roslyn Hole, Ely (see p.66). No specimen has survived but Jukes-Browne (1900, p.293) listed a number of fossils from the Gault and his identifications can be re-interpreted as including *Birostrina sulcata*, *Hysteroceras*, *Mortoniceras*, *Neohibolites minimus* and *Nielsenicrinus cretaceus*, an Upper Albian assemblage indicative of parts of the *cristatum* to *varicosum* subzones (Bed G 11 to G 14).

AAM, RWG

CHALK

The Chalk crops out in the most easterly part of the district where it forms a low escarpment that rises above Mildenhall and Methwold fens. Much of the outcrop is covered by Quaternary deposits and the sequence can be satisfactorily examined at the surface only between Methwold Common [696 937] and Blackdyke Farm [695 881], and between Holmsey Green [694 783] and Thistley Green [670 765]. The lowest part of the Chalk, including the junction with the Gault, is everywhere obscured, but these beds and the remainder of the Chalk that crops out in the district were proved in the cored boreholes for the Ely-Ouse Scheme between Blackdyke Farm and Holmsey Green. The generalised vertical section of the Lower Chalk of the Ely district, based on the sections proved in these boreholes, is shown in Figure 23.

The Chalk of the district is sparsely fossiliferous at most levels and has yielded only a few bivalves, brachiopods, ammonites, rare solitary corals, an isopod, bits of wood and trace fossils. All the fauna referred to in the following account has been determined by Mr C. J. Wood who has also commented on its zonal significance.

The arrival of the Chalk in the district heralded a long period of marine deposition in quiet, relatively deep water. Only the lower part of the Lower Chalk, a maximum of about 50 m of strata, is now preserved in the district but a complete Chalk sequence, probably 700 to 800 m thick, is likely to have been deposited and to have been largely removed by erosion in Tertiary times. There are marked thickness variations in the beds in the Chalk, even within the relatively short length of outcrop in the Ely district. The thickness of the Chalk below the Inoceramus Bed in the district remains relatively constant at 6.1 to 7.3 m, but that of the beds between the Inoceramus Bed and the Totternhoe Stone varies considerably. Erosion at the base of the last-named bed has locally removed large quantities of the underlying Chalk so that the stratigraphical distance between these two beds varies between 13 and 21 m within a few kilometres. The Chalk below the Totternhoe Stone (including 1 to 2 m of Cambridge Greensand) shows a general increase in thickness southwards from an estimated 25 m at Blackdyke Farm to 28.3 m at Holmsey Green, but within this distance these beds are locally as thin as 19 m where the Totternhoe Stone is especially erosive (Figure 24).

The Chalk of the Ely district lies close to the junction of the supposed 'northern' and 'southern' faunal and lithological provinces of the English Chalk. The Lower Chalk thins steadily northwards from the Cambridge district (typical 'southern province' rhythms of grey marly chalks and harder white chalks overlain by grey and white uniform chalks) to the Hunstanton district ('northern province' indurated, hard nodular chalks). However, a number of distinctive marker bands, notably the Totternhoe Stone and the Melbourn Rock, occur throughout both districts, and the supposed faunal and lithological differences at this stratigraphical level have been exaggerated in the past. The Lower Chalk of the Ely district shows some lithological characters that are similar to those in the 'northern province' and others that are more characteristic of the 'southern province'. The beds below the top of the Inoceramus Bed appear to be an expanded equivalent of the Paradoxica and Inoceramus beds (cf. Peake and Hancock, 1961) at Hunstanton: the Lower Chalk above the Inoceramus Bed is lithologically similar in the Cambridge, Ely and Hunstanton districts despite considerable variation in thickness.

The purity, fine-grained texture, unlithified nature and general homogeneity of the Chalk puzzled geologists for more than 150 years. The advent of the electron microscope enabled the fine detail of its structure to be resolved satisfactorily and it has become clear that it is composed largely of the remains of calcareous algae (Black, 1953). No new sedimentary work has been carried out on the Chalk during the present survey because its lithology and fauna and, by inference, its conditions of deposition, can be closely matched with that of adjacent areas. Wood (*in* Worssam and Taylor, 1969, pp.43–45) has described the sedimentology of the Chalk of the Cambridge district; Hancock (1975) has comprehensively reviewed the available data on the petrology and chemistry of the Chalk.

Chalk is composed largely of low-magnesium (less than 0.5 per cent Mg) calcium carbonate derived mostly from coccolithic algae (as disaggregated calcite plates, coccoliths and coccospheres), with foraminifera, bivalve shells,

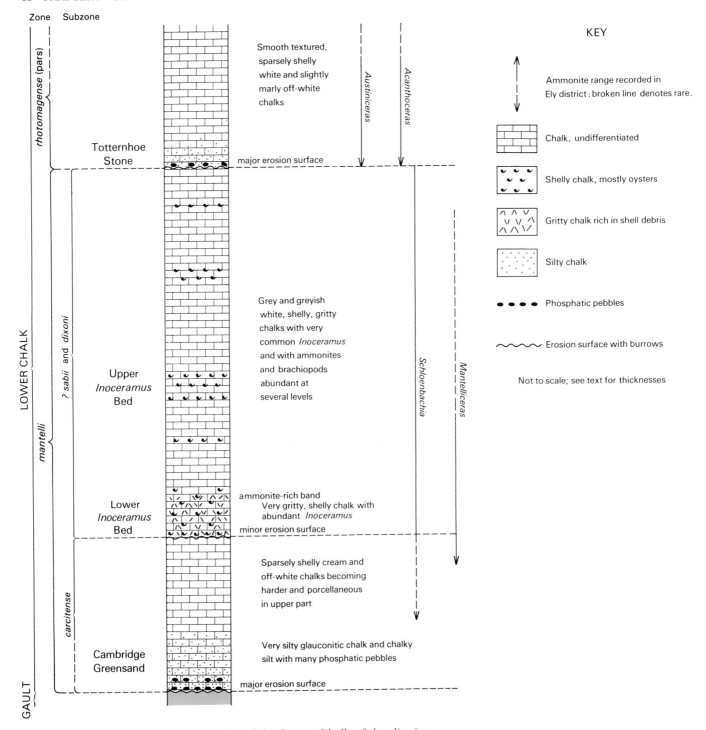

Figure 23   Generalised vertical section of the Lower Chalk of the district

calcispheres, echinoderm plates and bryozoan fragments forming important constituents at some levels. The clay-mineral content of the Chalk varies from less than 1 per cent in the purest chalks up to 30 per cent in thin marly layers. Jeans (1968) has described the clay minerals in the Lower Chalk of Cambridgeshire and Norfolk as composed largely of smectite and illite: recent work by Harrison and others (1979) suggests that this smectite may be partly volcanic in origin.

The fauna of the Chalk shows that it was deposited in a sea with normal salinity. Firm evidence for the depth of the sea is difficult to obtain, but Hancock (1975) suggested that the combined faunal and petrological data indicated that the purer chalks were deposited in about 100 to 600 m of water. Burnaby (1962) and Kennedy (1970) suggested that parts of the Lower Chalk were deposited at depths of less than 50 m on the basis of their foraminiferal and gastropodal contents respectively, but Hancock (1975) found this evidence unconvincing.

Parts of the Lower Chalk sequence of the Ely district, in particular the Cambridge Greensand, the Inoceramus Bed and the Totternhoe Stone, contain phosphatised chalk pebbles and silt- and sand-grade shell fragments in such quantities that they are not really chalks. Such beds were clearly either deposited at shallow depths in high energy environments or were formed in such environments and subsequently swept into deeper water. The widespread nature of these beds, the presence of common unbroken shells within them, and their lack of sorting or grading suggests that the latter possibility is unlikely. Furthermore, the unconformable relationship of the Cambridge Greensand to the Gault throughout the present district suggests that it was deposited in shallower, more current-agitated water than was the Upper Gault on which it rests. However, the Gault itself has been suggested to be a relatively shallow-water deposit (Gallois and Morter, 1982). A similar argument can be used to suggest a shallow-water origin for the Totternhoe Stone and, to a lesser extent, the Inoceramus Bed. The intervening chalks were deposited in quieter, and probably deeper, water below the influence of currents where there was little supply of siliciclastic or benthonic bioclastic materials.

The Chalk sea seems to have undergone rhythmic variations in depth, each rhythm beginning with a rapid shallowing which was followed by steady deepening. These and similar rhythms are so widespread in the Chalk of Europe that Hancock and Kauffman (1979) have argued that they result from world-wide eustatic changes in sea level. The steady overall upward diminution of siliciclastic material in the Lower Chalk probably reflects gradual erosion and submersion of the land areas that supplied these materials.

Cambridge Greensand

The Chalk rests unconformably on the Gault throughout the district. The unconformity is marked by an irregular burrowed erosion surface overlain by a distinctive bed of green-speckled glauconitic calcareous siltstone up to 0.3 m thick, crowded with green glauconite-coated phosphatic pebbles. This passes up into sparsely pebbly, silty and gritty chalk, about 1.7 m thick. In the Cambridge district, where it is thickest, this bed has been named the Cambridge Greensand. This unusual lithology was formerly well exposed in phosphate diggings between Soham and Cambridge where the bed contained a great variety of pebbles and boulders, and a rich phosphatised fauna in which ammonites and bivalves were abundant.

The most remarkable feature of the Cambridge Greensand is the presence within it of angular boulders of over 40 rock types, including igneous and high-grade metamorphic rocks, up to 55 cm across and weighing up to 60 kg. Hawkes (1943) concluded from a study of the petrology of these boulders that they were derived from the west because they include specimens of paisanite (a porphyritic riebeckite-microgranite), a distinctive rock whose only known outcrop in western Europe is at Mynydd Mawr, North Wales, and tuffs and rhyolites that can be matched with Precambrian Uriconian rocks from the Midlands and Welsh Borderlands. The method of transport to their present resting place in Cambridgeshire is believed to have been in the roots of floating trees. Although a number of authors have expressed scepticism as to how marine currents could concentrate such a great variety of rocks derived from such geographically dispersed sources into an area limited to about 25 km of the present-day Cambridge Greensand outcrop, no alternative explanation has been put forward. A nearby source to the south, east or north-east is unlikely because no igneous or metamorphic rock, and few arenaceous rocks, have been proved beneath the Cretaceous rocks of the London Platform. Furthermore, it is likely that the London Platform, and its south-easterly extension into the Ardennes of Belgium, was submerged at the time of deposition of the Cambridge Greensand. The large variety and number of igneous and high-grade metamorphic rocks found in the Cambridge Greensand also seems to preclude East Anglia as a source area. Such rocks must have been derived from an extensive igneous/metamorphic complex; the nearest such complexes to the district that are likely to have been exposed at this time are those suggested by Hawkes (1943). Until more is known about the concealed geology of the London Platform and its south-easterly extension and about the distribution of the Cambridge Greensand beneath the Chalk outcrop, the problem must remain unresolved. RWG

The age of the Cambridge Greensand cannot be directly demonstrated on the basis of its indigenous macrofauna; this is dominated by bivalves such as *Aucellina*, including *A. gryphaeoides* (J. de C. Sowerby), *A. krasnopolskii* (Pavlov) and *A. uerpmanni* Polutoff, *Entolium* and *Plicatula*. The remaining fauna includes the bivalves *Plagiostoma globosa* (J. de C. Sowerby), *Pseudolimea cantabrigensis* Woods, *Pycnodonte* spp. and rare inoceramid fragments, the brachiopods *Concinnithyris* spp., *Monticlarella carteri* (Davidson), '*Terebratula*' *biplicata* auctt. and *Terebratulina triangularis* Etheridge, the sponge *Pharetrospongia strahani* Sollas and the coral *Onchotrochus carteri* Duncan.

A rich and diverse derived ammonite and bivalve fauna was obtained from the former coprolite workings in the Cambridge and Ely districts, but all the ammonites are densely phosphatised and most show signs of wear. Parts of the non-ammonite fauna, mostly brachiopods and bivalves, are preserved in softer phosphate and enclose parts of the original shell; these are probably younger than the more densely phosphatised parts of the fauna, including the ammonites. Spath (*in* Osborne White, 1932) concluded that all the ammonites were late Albian in age, and that the deposit was formed in late Albian times by reworking of earlier deposits.

Breistroffer (1947a, 1947b, 1965) and Casey (*in* Edmonds and Dinham, 1965, p.55) both concluded that the phosphatised ammonite fauna in the Cambridge Greensand was late Albian (*dispar* Zone) in age and was derived from the Gault; Owen (1979, p.580) has pointed out that *auritus* Subzone elements are also present. The faunal content of the pebbles varies from place to place as does the age of the Gault on which the Cambridge Greensand rests. There is evidence in the Ely district, and in the adjacent Wisbech and Cambridge districts, to suggest that the Cambridge Greensand oversteps the Gault (Beds G 16 to G 18) in a southerly direction in that area. Faunas derived from Beds G 16 to G 18 (*Callihoplites auritus* and *Mortoniceras rostratum* subzones) have been recorded in the Cambridge Greensand together with a small amount of material secondarily derived from older subzones via the phosphatic pebble beds at the bases of Beds G 16 and G 17.

The Cambridge Greensand passes up without obvious sedimentological break into the chalky limestones of the Lower Chalk. These are rich in forms of the ammonites *Mantelliceras* and *Schloenbachia* that have generally been assigned to the Lower Cenomanian. Three specimens of *Schloenbachia* that were thought by Cookson and Hughes (1964) to have come from the matrix of the Cambridge Greensand indicated to them a Cenomanian age for the deposit. However, Casey (*in* Edmonds and Dinham, 1965), in a review of the age of the ammonite fauna of the Cambridge Greensand, concluded that there was no palaeontological evidence for a Cenomanian age because these specimens were unlikely to have been found in situ. He also noted that the stratigraphical range of *Schloenbachia* was itself in doubt pending more detailed work on the ammonite sequence of the sediments close to the Albian–Cenomanian boundary. Hart (1973) has assigned the Cambridge Greensand to the Lower Cenomanian on the basis of foraminifera. Here too, an element of uncertainty remains because of the lack of available detail of the ammonite sequences at the stage boundary in the sections used as the type for comparison of the foraminiferal sequences.   RWG, AAM

Lower Chalk between the Cambridge Greensand and the Totternhoe Stone

The Cambridge Greensand passes up into 6 to 7 m of hard cream and white, smooth textured, almost barren, in part porcellaneous chalks that are lithologically unlike the argillaceous chalks (Chalk Marl) at this stratigraphical level elsewhere in southern England, but are reminiscent of the basal bed of the Chalk, the Paradoxica Bed, of the Hunstanton district. These hard chalks are overlain by a distinctive bed, 2 to 3 m thick, of grey, very gritty (due to *Inoceramus* prisms), *Inoceramus*-rich grey chalk that is similar in lithology and fauna to the Inoceramus Bed of Hunstanton. Above this are up to 21 m of greyish white marly chalks which alternate with shelly and gritty chalks containing common *Inoceramus* and with ammonites and brachiopods abundant at a few levels.

The chalk between the Cambridge Greensand and the Inoceramus Bed in the Ely-Ouse boreholes contains ammonites indicative of an early Cenomanian age. These include species of *Idiohamites* and *Schloenbachia* that probably correlate with the ammonite assemblage of the *Neostlingoceras carcitanense* Assemblage Subzone of the *Mantelliceras mantelli* Zone of southern England (*sensu* Kennedy and Hancock, 1978). The sparse non-ammonite fauna comprises mainly the bivalves *Aucellina gryphaeoides* and *A. uerpmanni* and '*Inoceramus*', including '*I.*' cf. *anglicus conjugalis* Woods, with some terebratulid brachiopods.

The Inoceramus Bed, and a number of fossiliferous bands between it and the Totternhoe Stone, contain common *Schloenbachia*, and rare *Hypoturrilites*, *Mantelliceras* and *Scaphites obliquus* J. Sowerby, that correlate with the *Mantelliceras saxbii* Assemblage Subzone and probably with part of the *M. dixoni* Assemblage Subzone of southern England (*sensu* Kennedy and Hancock, 1978). An ammonite-rich band in the top part of the Inoceramus Bed seems to be a widespread marker in this and adjacent districts. The non-ammonite fauna is dominated by *Inoceramus*, both as fragmentary and whole shells, but includes rhynchonellid and terebratulid brachiopods at some levels.

The fauna of the Inoceramus Bed is composed mostly of '*Inoceramus*' ex gr. *crippsi* Mantell but includes some '*I.*' *reachensis* Etheridge and rare '*I.*' *anglicus conjugalis*. *Anomia sp.*, *Entolium sp.*, *Plicatula inflata* J. de C. Sowerby, rhynchonellids including *Grasirhynchia grasiana* (d'Orbigny) and large terebratulids are also present. A second, but less well developed *Inoceramus*-rich bed, forms a persistent, stratigraphically higher marker throughout the district, and is referred to in Figure 22 as the 'Upper' Inoceramus Bed. This bed, and the beds between it and the Totternhoe Stone, contain common *I.* ex gr. *virgatus* Schlüter and '*I.*' *reachensis* with subordinate '*I.*' *crippsi*. The age of the highest part of these beds is still in doubt. Rawson and others (1978, p. 50) have suggested, on the basis of specimens from the Cambridge district, that the *mantelli–rhotomagense* zonal boundary lies below the Totternhoe Stone in some sections. In areas such as the Ely district, where erosion at the base of the Totternhoe Stone can be shown to have locally removed large parts of the underlying sequence, this zonal boundary is probably in many places coincident with the erosion surface.

Totternhoe Stone and overlying Lower Chalk

The Totternhoe Stone is a bed of silty intensely bioturbated grey chalk, about 1.5 m thick, with pebbles of phosphatised and glauconitised chalk in its lower part, that rests erosively on the underlying beds and, in places, cuts out a considerable thickness of them (Plate 6). This bed is one of the most lithologically distinctive and persistent markers in the Chalk of East Anglia and its outcrop has been traced by means of exposures and a topographical feature in the Cambridge and Ely districts.

The Lower Chalk above the Totternhoe Stone is much whiter, less argillaceous, and more uniform in texture than the beds below, and mostly breaks with a subconchoidal fracture; up to 20 m are preserved. The highest beds of the Lower Chalk are not exposed in the district, but they, and the distinctive pebbly and nodular chalks of the overlying Melbourn Rock, crop out on a prominent feature capped by the Melbourn Rock that lies a few hundred metres east of the district boundary between Whitedyke Farm and Blackdyke Farm, Hockwold.

The Totternhoe Stone at Blackdyke Farm has yielded a fauna indicative of the *rhotomagense* Zone including the large ammonites *Acanthoceras* and *Austiniceras*.

DETAILS

Ely-Ouse boreholes

The whole of the Chalk, from its base to a level a little above the Totternhoe Stone, was proved in the boreholes drilled for the Ely-Ouse Water Transfer Scheme between Blackdyke Farm [6916 8817] and Holmsey Green [6962 7858]. The following sequence is that of Borehole 15 [6949 8213] which proved the most complete section: similar sequences were proved in Boreholes 12 to 23. The positions of the main marker bands in the better documented of these boreholes are shown in Figure 24.

	Thickness m	Depth m
Base of Recent deposits	c.3.7	c.3.7

Acanthoceras rhotomagense Zone
Chalk, white with patchy yellow limonitic
staining; smooth textured with clean fracture;
sparsely shelly with rare *Inoceramus* chips and
fish debris; burrows and trails including rare
glauconite-lined burrowfills; passing down into   c.4.8   8.45

TOTTERNHOE STONE: intensely bioturbated,
mottled grey, brownish grey and creamy
white, gritty (shell debris) and silty chalks;
becoming more gritty and silty with depth;
sparsely shelly with *Inoceramus* fragments the
only common fossil; becoming pebbly below
8.8 m with pebbles of phosphatised (cream coloured) and densely phosphatised and
glauconitised (brown with green rims) chalk
up to 5 cm across; all pebbles bored, the more
densely phosphatised having at least two
generations of boring; very irregular and com-
plexly burrowed junction with bed below   1.12   9.57

Mantelliceras mantelli Zone
Mantelliceras dixoni ? and *M. saxbii* subzones
Chalk, off white and slightly greyish white
with patchy, pale yellow limonitic staining;
gritty textured throughout; large burrowfills of
Totternhoe Stone common down to 9.90 m,
small burrowfills common down to 10.21 m
with rare burrowfills down to 10.81 m; sparsely shelly but becoming shelly with depth with
much *Inoceramus* debris and rare complete
shells; small brachiopods, including rhynchonellids and *Terebratulina*, common at several
levels, notably at 10.66 and 10.97 m; concentration of *Inoceramus* forms an 'Upper' Inoceramus Bed at 14.50 to 15.01 m; rare
sponges, trails and burrows occur throughout;
passing down into   10.39   19.96

INOCERAMUS BED: bioturbated, tough, gritty,
grey chalk with much *Inoceramus* debris and
common whole shells; ammonites, mostly
Schloenbachia, common at several levels,
especially in top part of bed, some limonite-
and/or glauconite-coated; nests of shell debris
including rhynchonellids; burrowed basal junction   2.59   22.55

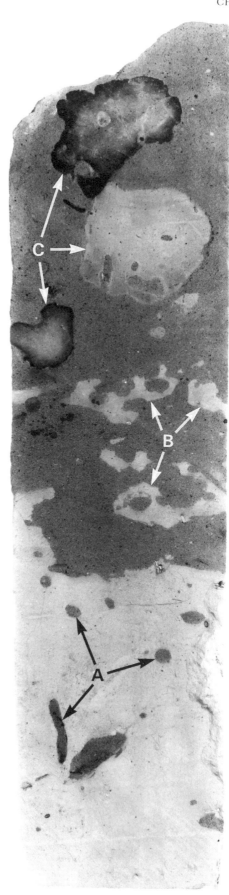

Plate 6 Structures in the basal bed of the Totternhoe
Stone, Ely-Ouse Borehole 15. The grey gritty matrix of
the Totternhoe Stone contrasts with the underlying white
chalk. Burrow infillings of Totternhoe Stone penetrate the
white chalk (A), and vice versa (B). Pebbles of green,
glauconite-coated, partially phosphatised and calcitised
chalk (C), many with bored outer surfaces, are common
within the Totternhoe Stone (BDR 6329)

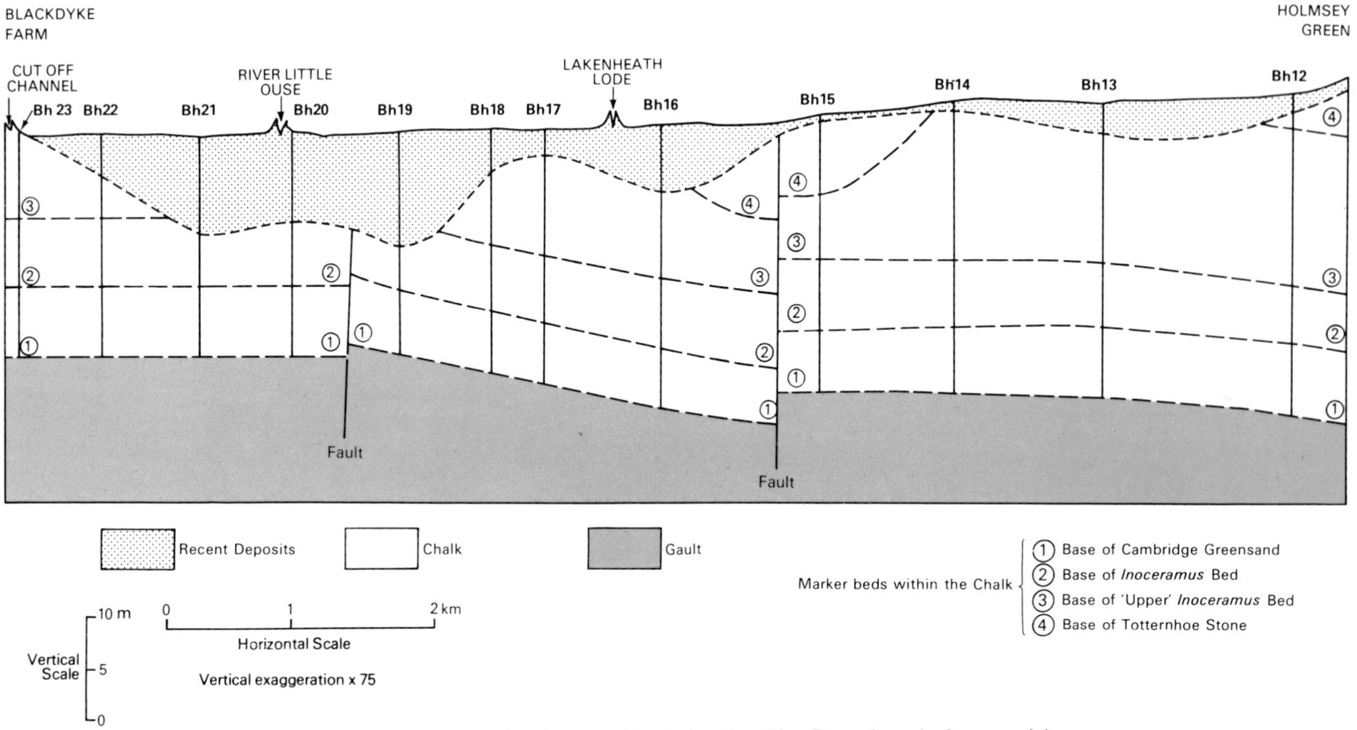

Figure 24   Correlation of marker bands in the Lower Chalk in the Ely-Ouse boreholes: positions of faults after Samuels (1975)

	Thickness m	Depth m

?Neostlingoceras carcitanense Subzone
Chalk, uniform creamy white, smooth textured, hard and porcellaneous at several levels in upper part of bed; rare, thin marl wisps; sparsely shelly but with varied fauna including *Aucellina*, *Inoceramus*, and other bivalves; ammonites including *Idiohamites* cf. *alternatus* (Mantell) and *Schloenbachia* spp., rhynchonellid and terebratulid brachiopods, sponges, fish debris and burrows and trails; these last commonly lined with pale green glauconitised clay; *Aucellina* the only common fossil in lowest part of bed; passing down into 3.96   26.51
CAMBRIDGE GREENSAND: pale grey, marly and silty chalk passing rapidly down into medium slightly greenish grey clayey and chalky silt with rare small, black (very dark green coated) phosphatic pebbles; sparsely shelly with *Aucellina* spp. the only common fossil; base in core loss c.2.2   c.28.7

Hockwold area
Chalk was formerly worked in small pits near Whitedyke Farm [693 896 and 692 890], Blackdyke Farm [690 886], Kennyhill [668 796]

and Thistley Green [679 767] all at levels close to the Totternhoe Stone. A specimen collected by Dixon Hewitt from Whitedyke Farm, presumably from one of the two pits listed above, was determined by Peake and Hancock (1961, p.301) as *Calycoceras*, but has been redetermined by Mr Wood as probably a crushed *Mantelliceras*. The upper part of the Blackdyke Farm section is still exposed, but the remaining pits are now degraded. Peake and Hancock (1961, p.301) recorded Chalk with abundant *Schloenbachia*, including *S. varians* (J. Sowerby), *S. intermedia* (Mantell) and *S. subvarians* (Spath), overlain by Totternhoe Stone with common *Austiniceras* at Blackdyke Farm. Only the upper part of the Totternhoe Stone is now exposed.

The Cut-Off Channel between Whitedyke and Blackdyke farms is mostly excavated in Chalk at a level a little below the Totternhoe Stone. The sections were deeply weathered and yielded little determinable faunal material.

Holmsey Green
Fossiliferous bands within the Chalk have been exposed from time to time in the deeper drains on the Chalk outcrop or where the superficial deposits are thin. Dr C.R. Bristow collected fossiliferous chalk from beneath terrace deposits in a ditch section at Hicks' House [6925 7974], near Peterhouse Farm. The fauna has been determined by Mr Wood as *Anomia sp.*, '*Inoceramus*' ex gr. *crippsi*, '*I*'. *scalprum* Böhm, *Plagiostoma globosa* (J. de C. Sowerby) and *Schloenbachia varians*, and probably comes from the beds between the Inoceramus Bed and the Totternhoe Stone.   RWG

CHAPTER 6

Quaternary deposits

Quaternary deposits crop out over more than four fifths of the Ely district. They can be divided into two on the basis of age; those that were deposited during the glacial and interglacial episodes of the latter part of the Pleistocene period some 150 000 to 10 000 years ago, and those that have accumulated in a temperate climate during the last 10 000 years. The younger deposits, the Recent (Flandrian) marine and freshwater deposits that form Fenland, have by far the greater area of outcrop in the district.

PLEISTOCENE

Chronology

The present distribution of Pleistocene deposits in the district is patchy (Figure 25) and provides only a fragmentary record of the history of the period. The succession and history deduced for the district, together with tentative correlations with adjacent districts, are shown in Table 5.

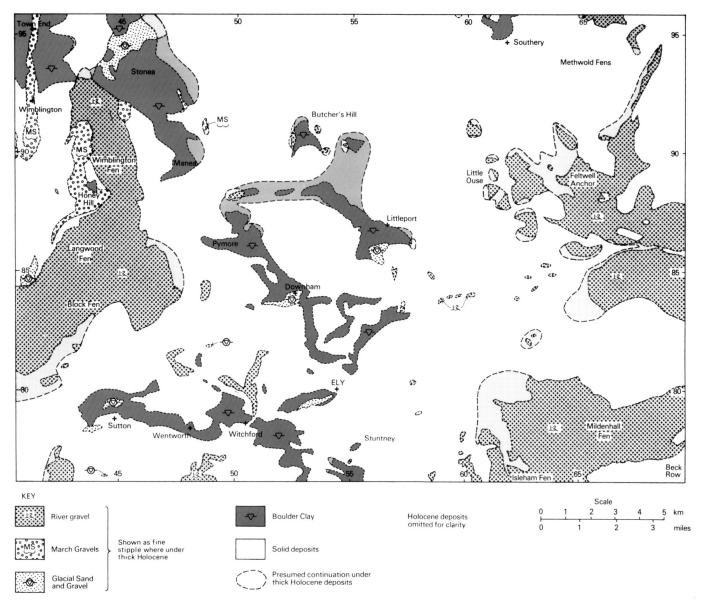

KEY

River gravel · March Gravels · Glacial Sand and Gravel — Shown as fine stipple where under thick Holocene

Boulder Clay

Solid deposits

Presumed continuation under thick Holocene deposits

Holocene deposits omitted for clarity

Scale

Figure 25   Distribution of the Pleistocene deposits of the Ely district

Table 5 Correlation of the Pleistocene deposits of the Ely district with those of adjacent districts

Deposits			Erosional features in Fenland region	Approximate age in years before present	Presumed sea level relative to present
The Wash and north-eastern Fenland	Ely district	Cambridge district			
Solifluction deposits	Solifluction deposits; formation of cryoturbation and ground-ice features		Meltwater erosion of solid and older Pleisto-cene deposits	10 000 to ~15 000	Rising from −100 to −30 m
Hunstanton till and associated sand and gravel				~15 000 to 25 000	−100 m
Wretton Terrace of Wissey	First and Second terraces of Great Ouse, Little Ouse and Lark	First and Second terraces of Cam and Lark	?Formation of Fenland 'islands'	~25 000 to >46 000	slightly lower than present
Hunstanton raised beach; ?Upper Tottenhill Gravels	March Gravels	Third and Fourth terraces of Cam and Lark	Meltwater erosion of older deposits	>46 000	+3 to +5 m
Lower Tottenhill Gravels	—	Head Gravel and Observatory Gravels	Extensive erosion of older Pleistocene deposits		lower than present
Nar Valley Clay	? Salt marsh clay at March	—	—		Rising from −8.5 to +30 m
Nar Valley Freshwater Beds	—	—	—		Rising from −20 to −8.5 m
Varved clays passing down into Chalky–Jurassic till with associated sand and gravel			Retreat of chalk escarp-ment; for-mation of tunnel valleys; modification of fluvial valleys		−100 m
Pre-glacial weathering products			Pre-glacial valley system		Falling from 0 to −100 m

The Pleistocene history of East Anglia has been a subject of controversy for more than 100 years, largely because of the inconclusive nature of much of the evidence. Part of the problem arises from the absence of a reliable method for determining the ages of Pleistocene deposits more than about 46 000 years old (the practical limit for radiocarbon dating), for though Szabo and Collins (1975) used a uranium method to determine the ages of bones in the range 100 000 to 250 000 years, this technique has not been widely used. Moreover, few Pleistocene deposits contain indigeneous shelly faunas, and those that do, have assemblages made up of long-ranging forms that are of little stratigraphical value. Similarly, the vertebrate faunas that are relatively common in some Pleistocene deposits, although commonly indicative of specific climatic conditions, are rarely stratigraphically useful. The most determined attempts to provide a palaeontologically-based stratigraphical framework for the Pleistocene have been based on pollen assemblages. Although many individual forms are long-ranging it has been claimed that the pollen spectra of deposits formed during particular warm episodes can be distinguished from one another. Attempts have also been made to use mammals, bivalves, gastropods, foraminifers, ostracods and beetles to construct climatic 'curves' in the hope that these will prove to be distinctive.

The knowledge that Pleistocene history is one of repetition of cold and warm climatic cycles led to the proposal (Mitchell and others, 1973) that the Pleistocene should be divided into cold and warm phases, the deposits of each phase being regarded as belonging to a climatic 'Stage' and allocated a name on the basis of a type section. This use of the term stage is now recognised to be unsatisfactory (Mitchell *in* Holland and others, 1978) because of the absence in Britain of sufficient palaeontological or other evidence to enable the

Pleistocene to be divided into chronozones and true stages. The proposed system of subdivision suffers, therefore, from the limitation that the individual stages cannot be uniquely defined. In practice, the application of the proposed scheme depends partly on the recognition of field evidence to enable the stratigraphical relationships of the different deposits to be determined, and partly on the assumption (based on the consensus view in 1973) that three glacial (Anglian, Wolstonian and Devensian) and two interglacial (Hoxnian and Ipswichian) phases are recognisable in Britain. These are collectively referred to as the British Standard Stages.

Work on the oxygen isotope ratios of air bubbles trapped in ice in the Greenland ice-cap (Dansgaard and others, 1971) and the ratios for calcium carbonates from deep-sea foraminiferal oozes (Shackleton and Opdyke, 1973) has shown however, that the Pleistocene period was characterised by many more climatic fluctuations than this classification allows. Similar isotopic data has been interpreted by Imbrie and others (1984) to indicate nine major cold phases, in which the global volume of ice was sufficient to depress sea-level by at least 100 m, during the past 800 000 years. In addition, each cold phase may have within it short warmer periods (interstadials), and interglacial periods may similarly contain short cold periods. To add to the difficulty of deducing the climatic history from the Pleistocene deposits, not all cold phases gave rise to extensive glaciations in Britain. The recognition of the British Standard Stages is, therefore, by no means simple or universally accepted.

The resultant confusion has led to doubt about the regional correlation of the Pleistocene deposits of Fenland. Thus Mitchell and others (1973) considered that the Chalky-Jurassic till[1] (Chalky Boulder Clay of some authors) of eastern Fenland falls within the Anglian Stage, but placed outcrops of what appear on lithology and provenance to be the same till in western Fenland (1973, Table 3) in the Wolstonian.

Elsewhere in East Anglia, Bristow and Cox (1973) suggested that deposits that had been assigned to the Anglian and Wolstonian stages were either of the same age or resulted from two cold phases within a single stage. Straw (1979) placed both the eastern and western outcrops of the Chalky-Jurassic till of Fenland within the Wolstonian. In doing this he suggested that an earlier (Anglian) ice-sheet covered the area, but that its deposits were removed prior to and during the advance of the Wolstonian ice. In the Ely district, only the Chalky-Jurassic till provides direct evidence that the area was covered by an ice-sheet. However, in adjacent areas of East Anglia, where extensive outcrops of till cover much of Norfolk, Suffolk and Cambridgeshire, Straw's interpretation (1979, fig.1) requires a junction to be drawn between two tills of apparently markedly differing ages in areas where they have yet to be shown to be lithologically and/or geomorphologically distinguishable.

In the light of these difficulties, the scanty evidence yielded by the deposits of the Ely district does not provide a positive contribution to the resolution of the classification of the Pleistocene of East Anglia.

1 The term till is used in this account to describe all the deposits formed as the ground moraine of a continental ice-sheet. The bulk of this material is shown on the map as Boulder Clay: it includes within it pockets of sand and gravel of varying sizes. Where these latter are sufficiently large they are shown on the map as Glacial Sand and Gravel.

BOULDER CLAY AND GLACIAL SAND AND GRAVEL

The oldest Pleistocene deposits in the Ely district are the Chalky-Jurassic till and the glacial sand and gravel associated with it. The till is restricted to the area west of a line running from Southery to Ely (Figure 25): a small outcrop of boulder clay at Shippea Hill, to the east of this line, has subsequently been shown by drilling to be a cryoturbation deposit. The present-day patches of till are probably the remnants of a sheet that was once almost continous across the district. The till fills a topography cut into the solid deposits and is composed almost entirely of Upper Jurassic and Cretaceous rocks derived from Cambridgeshire, Lincolnshire and Norfolk. Thin patches of these glacial deposits cap the ridges at Downham, Ely, Littleport, Southery, Sutton, Wimblington and Witchford, and form the Fenland 'islands' at Apes Hall, Butcher's Hill, Manea and Stonea.

Field evidence and borehole data suggest that the Chalky-Jurassic till and its sand and gravel are less than 10 m thick everywhere in the district except for a small area at March. There, a borehole [4179 9482] at Wimblington Road, Town End, proved 27 m of Chalky-Jurassic till. Neither the shape nor the extent of this thicker till sequence is known, but it has been suggested, from a consideration of the subglacial surface of Fenland and The Wash, that it infills a small valley made by a tributary of the pre-glacial river Great Ouse (Gallois 1979a, fig.17).

The Chalky-Jurassic till of the Ely district is characteristically composed of stiff, slightly sandy clay with fine and medium gravel-sized erratics of chalk, flint and Jurassic cementstones, and with rarer oolitic limestone and ironstone, Triassic marls and a few farther-travelled erratics including very rare igneous and metamorphic rocks. Most of the matrix has been derived from the Upper Jurassic clays of Fenland, and still contains recognisable fragments of material such as cementstones from the West Walton Beds, Ampthill Clay and Kimmeridge Clay, coccolith-rich limestones from the Kimmeridge Clay, thick-shelled oysters from the West Walton Beds, and oil shales from the Kimmeridge Clay.

The erratics suggest that the till was derived from the north-west or north, from beneath northern Fenland, The Wash, and the broad clay vale of Upper Jurassic rocks that lies between the escarpments of the Lincolnshire Limestone and the Chalk in Lincolnshire. A more westerly source would have included more Triassic and Lower and Middle Jurassic rocks, and a more easterly source could not have provided the Upper Jurassic rocks. The local abundance of Triassic rocks in the basal parts of some sections may indicate variations in the direction of ice movement or layers within the ice that had differing sources. Throughout most of the district, however, the composition of the preserved Chalky-Jurassic till is largely controlled by the local geology.

Unusually large erratics have been recorded from time to time in the Chalky-Jurassic till in East Anglia. One of the most famous of these occurs at Roslyn Hole, Ely [555 808], and was the subject of much controversy in Victorian times. Skertchly (1877, Plate XXIII) described the erratic as being about 50 m by over 400 m in area and more than 5 m thick, and composed of a more or less comfortable sequence of

Kimmeridge Clay, Lower Greensand [Woburn Sands], Gault, Upper Greensand [Cambridge Greensand] and Chalk (Figure 26). The erratic was first described by Sedgwick (1846), and he and several later workers, notably Seeley (1865a, 1865b, 1868), argued that these Cretaceous strata were *in situ*, having been brought into contact with the Kimmeridge Clay by a complex pattern of faults. Fisher (1868) and Bonney (1875), on the other hand preferred a glacial origin. The discussion was terminated by Skertchly (1877, p.236) who recorded the presence of boulder clay on every side of and beneath the erratic mass. The stratigraphical sequence in the erratic suggests a local origin. The Woburn Sands and Cambridge Greensand are absent north of the Little Ouse because of unconformity and lateral facies variation respectively. The erratic, transported by ice moving from either the west or north-west must, therefore, have been derived from a Cretaceous outlier or escarpment lying in the Ely-Littleport area or to the west of it. The lower beds within the erratic, the Kimmeridge Clay and Woburn Sands, still crop out on the Ely ridge, and the most likely explanation is that the erratic was derived from a Chalk escarpment that lay close to Roslyn Hole. The erratic and the enclosing boulder clay were probably deposited in a small valley, of either fluvial or fluvioglacial origin, as a 'leeside' till when the escarpment was overridden by the ice.

Large erratics of Cretaceous rocks are also known from the eastern edge of Fenland at Downham Market, Leziate and King's Lynn in Norfolk, and it has been suggested that these are close to the line of a pre-glacial Chalk escarpment that lay to the west of the present one. The roots of this escarpment are now marked in the Ely district by the 'islands' of Southery, Littleport and Ely (Gallois, 1979b).

Between the north Norfolk coast and Stoke Ferry, Norfolk the present day Chalk escarpment, although divided into two (a Lower Chalk and an Upper Chalk escarpment) and much lower than its southern counterparts in the Chilterns and in Kent and Sussex, lies within a few kilometres of its pre-glacial position. It was probably eroded by ice moving from the north-west at an oblique angle to it. Between Stoke Ferry and Ely the former escarpment appears to have been eroded by ice from the west that moved on to it at a more acute angle, causing the escarpment to be eroded back so that it now lies about 10 km east of its former position. This has given rise to the large embayment occupied by Methwold Fens, Feltwell Anchor, Burnt Fen, Great Fen and Isleham Fen. No glacial deposit has been recorded from this area, and it forms a drift-free zone separating the patchily drift-covered areas of the Jurassic outcrop from the completely drift-covered till plateau of Norfolk and Suffolk. Debris from the former escarpment is now scattered throughout the Chalky-Jurassic till of East Anglia.

Gravelly sands composed of flint and the other more resistant erratics from the Chalky-Jurassic till occur in association with the boulder clay at Downham, Ely, Littleport, Stonea and Sutton. They were formerly worked on a small scale for road-making and for use as ballast in the construction of the railways across Fenland, but are now poorly exposed and few details of their lithology are known. Their relationship to the Chalky-Jurassic till is complex. The gravels occur as irregular masses within, beneath and on top of the till (although this last feature may be a result of

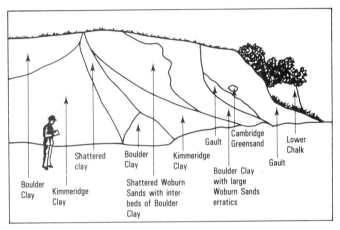

Figure 26 The 'Great Erratic' at Roslyn Hole, Ely in 1875 (a contemporary woodcut by S. B. J. Skertchley)

removal of a once overlying till), and can have either sharp or gradational contacts with the till.

On the lower ground around Stonea and Manea the upper part of the Chalky-Jurassic till locally contains lenses of interlaminated 'varved' clay and silt. Similar clays occur elsewhere in Fenland in the highest part of the Chalky-Jurassic till sequence where they were probably deposited in small, shallow, ice-dammed lakes during the latter part of the glaciation (Gallois, 1979b).

DETAILS

There is no permanent exposure of either Boulder Clay or Glacial Sand and Gravel within the district. Both materials have been worked for construction (embankments and roads) purposes in the past, and small temporary exposures have occurred in drains and ditches across the more poorly drained parts of their outcrops. The more resistant erratics commonly occur as plough debris and the lithology of the deposits can be deduced with reasonable accuracy in many areas, even where they are not exposed.

March to Chatteris

A borehole [4179 9482] at Wimblington Road, March, proved 27.0 m of Chalky-Jurassic till. The upper 20 m consists of chalk-rich clays, and the lowest 1 m of gravelly sand clay with much Triassic and Jurassic debris. A nearby borehole at the Auction Ground [4164 9525] penetrated 2.5 m of the till beneath 7.5 m of younger Pleistocene deposits.

A borehole [4506 9449] at White Gate Farm, Stonea, proved 1.75 m of cryoturbated orange-brown gravelly sand and chalky clay till overlying 3.55 m of interbedded chalky till, gravelly sand and varved clay resting on Ampthill Clay. The gravelly 'island' of Stonea was formerly interpreted as composed largely of March Gravels (see p.69 for details), but the borehole and adjacent drain sections indicate that the 'island' is mostly underlain by Chalky-Jurassic till and gravels that form a single glacial complex. Parts of the till sequence are exposed from time to time in the deep drains that cross the 'island', notably the Sixteen Foot Drain. Sections recorded in the banks of this drain at Stonebridge Farm [463 939] during the present survey showed up to 3 m of gravelly, sandy clay till with abundant small chalk and flint erratics: various Upper Jurassic lithologies and fossils have been collected from the till in this area in the past.

A borehole [4135 8487] at Wenny Farm, Chatteris, proved 4.0 m of cryoturbated gravelly sand passing down into sand and gravel with thin beds of varved clay. The gravels were originally thought to be March Gravels, but their stone content is similar to that of the Glacial Sand and Gravel at Stonea, and they contain no shell material other than fragments of thick-shelled oysters derived from the Jurassic.

Ely to Southery

Glacial Sand and Gravel was worked in a large borrow pit at Downham [528 842] for use as embanking material on the Ely to March railway: the composition of the larger stones in the gravel is probably reflected by the flint, Woburn Sands, brown quartzite pebbles (presumed from the Trias) and Jurassic cementstones in the walls of the adjacent church. A mixture of Chalky-Jurassic till and Kimmeridge Clay was also used on the railway and much of this came from a widened cutting at Chettisham [547 838]. A mixture of Kimmeridge Clay and Chalky-Jurassic till was also dug (for repairs to river banks) at Roslyn Hole, Ely [557 807], although in later years the till was worked out and only the Kimmeridge Clay is still in use.

A patch of Glacial Sand and Gravel gives rise to gravelly soils on the higher part of the 'island' of Littleport. A shallow degraded pit in this gravel at Mill Lane [5649 8634] may have worked flints for use in corn grinding.

Chalky-Jurassic till was exposed in a shallow cutting for the Southery Bypass [618 949] where it consisted of chalk-rich sandy clay, rich in glauconitic sand and phosphatic pebbles derived from the Sandringham Sands. Fitton (1836, p.316) recorded the Jurassic fossils *Deltoideum* and *Laevaptychus* from the till of this area.

A borehole [6158 8384] sited on the patch of clayey drift (presumed to be boulder clay) that caps the ridge at Shippea Hill Farm proved 0.5 m of stony sandy soil and subsoil resting on 1.6 m of cryoturbated gravelly sand and weathered Kimmeridge Clay. The sand contains pebbles of flint, quartzite (presumed Triassic), chalk, vein-quartz and lydite, and is presumed to be a remanié of a glacial sand and gravel deposit or a terrace deposit that has either been weathered in situ or derived by solifluction.

MARCH GRAVELS AND RIVER TERRACE DEPOSITS

The glaciation that deposited the Chalky-Jurassic till was followed by a temperate phase, probably an interglacial, during which the Ely district was presumably covered by the sea. No deposit of this age has been positively identified in the district but in the adjacent Wisbech district (Sheet 159) this phase is represented by the Nar Valley Beds, over 30 m of fluviatile sands and marine clays (Table 6). Only one possible representative of this phase has been recorded in the Ely district. At March, a borehole [4164 9525] at the Auction Ground, Town End, proved 0.5 m of pale grey, soft silty clay with *Hydrobia* and peat rootlets, that is lithologically similar to some of the salt-marsh and intertidal clays forming at the present day in the southern part of The Wash. These clays overlie about 1 m of gravelly sand which rests unconformably on the Chalky-Jurassic till; they are themselves unconformably overlain by the March Gravels (see below). In the Nar Valley the lower (freshwater) part of the Nar Valley Beds are transgressed and overstepped by the upper (marine) part, and the intertidal clay at Town End may have been deposited during this transgression. However, because faunal or floral evidence is absent the Town End deposit could have formed during any of several temperate phases that may have occurred between the retreat of the 'Chalky-Jurassic' ice and the deposition of the March Gravels.

In the Wisbech district the Nar Valley Beds are overlain by coarse, torrential gravels (the Lower Tottenhill Gravels) that contain much debris derived from the Nar Valley Beds and the Chalky-Jurassic till. In the Cambridge district coarse gravels classified as Head Gravel and Observatory Gravels (Worssam and Taylor, 1969) have a field relationship to the Chalky-Jurassic till similar to that of the Lower Tottenhill Gravels, and both may have been deposited during the cold phase that followed the deposition of the Nar Valley Beds. There is as yet no palaeontological evidence to support their correlation.

At March, Wimblington, Honey Hill and near Stonea the Chalky-Jurassic till is unconformably overlain by up to 6 m of shelly, cross-bedded, sand and flint gravel that Baden Powell (1934) termed the March Gravels. These gravels were formerly exposed in a large number of small pits where they were worked for aggregate or ballast, but these are now overgrown and any discussion of the age and origin of the March Gravels must rely heavily on earlier accounts and fossil collections. The fauna of the gravels is dominated by species of *Macoma*, especially *M. balthica* (Linnaeus): all these shells are slightly abraded and some are broken, but most are well preserved, whole, and have retained some of their original colour banding. There is little doubt that they are indigenous. The remaining fauna includes species of *Cardium*, *Cerastoderma*, *Corbicula*, *Mytilus*, *Ostrea* and *Tellina*; *Turritella* is locally abundant and pebbles bored by *Pholas* are present at some localities. Rare scaphopods, brachiopods and echinoid spines have also been recorded. The fauna indicates deposition in a cool, temperate marine environment and the great majority of the species recorded inhabit The Wash at the present day. Baden Powell (1934) noted that one specimen, a single valve in the Sedgwick Museum attributed to *Tellina obliqua* J. Sowerby, appeared to be from an extinct species. This species was also recorded from the Nar Valley Beds and from a raised beach at Hunstanton, and was used by Baden Powell (1934) to correlate these three deposits. Subsequent field work (Gallois, 1979b) has shown that the Nar Valley Beds are older than the Hunstanton Raised Beach (see below).

Baden Powell (1934, p.195) believed that the marine shells in the gravels were confined to heights between OD and 6 m above OD, and suggested that this indicated deposition in an intertidal zone at a time when mean sea-level was about 7 m

Table 6 Summary of the Flandrian history of the Ely district

Thousands of years BP	Formation in Ely district	Transgressions in Ely district	Transgressions on Dutch and French coasts	Archaeological period	Pollen zone
0					
	Period of reclamation	?Medieval flooding ▶	Dunkerke III ▶	Modern	VIII
1	– – – –			Medieval	
	Terrington Beds	Roman flooding ▶	Dunkerke II ▶		
2				Roman	
		Iron age flooding ▶	Dunkerke I ▶	Iron Age	
3	Nordelph Peat*				VIIb
			Dunkerke 0 ▶	Bronze Age	
4	Barroway Drove Beds				
		▶	Calais IV ▶	Neolithic	
5			Calais III ▶		VIIa
	'Lower' Peat				
6			Calais II ▶		
7	– – – –				
	'Lower' Peat in hollows and river channels			Mesolithic	VI
8			Calais I ▶		
	– – – –				
9	? Basal sand and gravel				V
10					IV

* In those parts of the Fenland not reached by the Terrington Beds transgression, the Nordelph Peat continued to form until it was drained and reclaimed during the Medieval period.

higher than at the present time. However, the gravels do not form a single continuous spread within the topographical interval described by Baden Powell, but occur as discrete patches at differing heights. The upper surface of the strip of gravel that caps the March-Wimblington ridge is mostly at 4 to 6 m above OD; that of the gravels that crop out between Honey Hill and Wimblington Common falls gently northwards from 4 m above OD to just below OD. Small patches of gravel described as shelly by Baden Powell (1934) at Cow Common [490 912], The Dams (480 923) and Jenny Gray's Farm [455 917] are a little above OD.

The age and mode of formation of the March Gravels is still not entirely clear. There is undisputed evidence that the March Gravels are younger than the Chalky-Jurassic till; sections showing the gravel overlying the till were described by Skertchly (1877), Whitaker and others (1891) and Baden Powell (1934). Two closely spaced boreholes [4179 9482 and 4164 9525] drilled at about the same height above OD at Town End, March, proved respectively the till at the surface, and 5.6 m of March Gravels on the till. They seem to indicate that the gravels are banked up against a low cliff or steep slope cut in the till (Gallois, 1976).

The March Gravels that cap the March-Wimblington ridge and Honey Hill appear to pre-date the formation of the valley in which the terrace gravels of Block Fen (see below) were deposited. In the Wimblington and Honey Hill areas the two deposits are separated by valley slopes cut in Ampthill Clay (Figure 27). At Wimblington Common, the gentle northward slope of the March Gravels brings the Honey Hill outcrop down to the level of the terrace gravels and into contact with them.

There is circumstantial evidence to suggest that the Hunstanton Raised Beach of north west Norfolk and the March Gravels are of the same age. Their lithologies differ, but their faunas and heights above O.D. are similar to one another, and indicate marine deposition in a temperate climate when the sea level was a little above that at present. However, such evidence needs to be treated with caution because of the repetitive nature of the climatic events that occurred in the Pleistocene. Many of the temperate phases, whether interglacial or interstadial, probably produced sea-levels comparable to or higher than those of the present day, and each maximum level is likely to have left behind patches of its more resistant deposits, such as gravels. As yet, there is no sure way of distinguishing the ages of such deposits: correlations made on the basis of their heights above sea-level are valid only on the assumption that they represent the deposits of a single temperate phase.

The March Gravels were believed by Skertchly (1877, p.201) to have been deposited as a marine beach and by Baden Powell (1934) as 'sand-banks' in a marine bay during an interglacial phase. The sedimentary features within the gravels, do not, however, confirm these suggestions. Firstly, the included flints are mostly angular and poorly sorted, and are unlike the well-sorted and well rounded flints of the modern storm beach that runs along the eastern margin of The Wash from Hunstanton to Snettisham. Secondly, in the few sections where it is still visible, the bedding in the March Gravels is mostly in thin units within which the cross-bedding dips northwards (i.e. towards The Wash); the style of bedding is similar to that produced in fluvial floodplains or fans and unlike that of shingle fulls developed on storm beaches. In addition, the geographical distribution of the March Gravels is difficult to explain if they are beach gravels.

When their composition, sedimentary features and distribution are compared with those of the river terrace gravels of the district (see below), the only major difference is the presence of marine shells in the March Gravels. Seale (1975) noted that the two gravels are impossible to distinguish in some parts of the Ely district, and that Baden Powell's (1934) differentation on the basis of their position relative to Ordnance Datum is unreliable.

The low-lying outcrops of March Gravels at Wimblington Common and The Dams could be interpreted as shelly patches of river terrace. By analogy, the March Gravels of the March-Wimblington ridge (the type section for the deposit) could be regarded as a fan deposited in a shallow marine bay into which the rivers that deposited the higher terraces debouched. This origin for the March Gravels would explain the source of the flints and other stones and their cross-bedding features, and would be in accord with their angularity and lack of sorting. The marine environment envisaged can be compared with the southern part of the present-day Wash, one of low energy protected from all but north-easterly storms. In such an environment, rounding

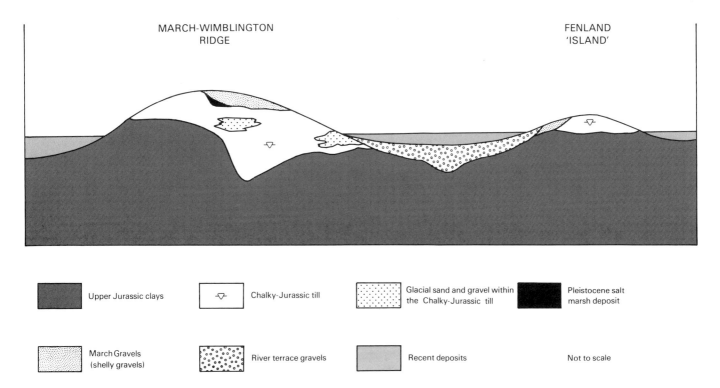

Figure 27   Diagrammatic section showing the stratigraphical relationships of the Pleistocene deposits in the north-western part of the district

and re-sorting of the gravels would be minimal, while shallow water burrowing bivalves, such as *Cerastoderma*, *Macoma* and *Tellina*, would be attracted to the sandy and more finely gravelly parts of the fan and many would be preserved with unbroken shells. The presence of the brackish or freshwater bivalve *Corbula fluminalis* (Müller), the freshwater *Unio tumidus* Retzius and freshwater species of the gastropods *Bithinia*, *Limnaea*, *Pisidium* and *Planorbis*, all recorded by Baden Powell (1934), would be explained, as would the presence of halophytic plants in the nearby terrace gravels at Earith (see below).

Thus, although some long-shore drift may have occurred and local beaches might have been formed, it seems likely that the bulk of the March Gravels are the seaward extension of river terrace gravels of the Great Ouse and Cam as implied by Skertchly (1877, pl. 18). If this is the case then pockets of shelly gravel are likely to be present in the most seaward parts of many of the spreads of 'Fen Edge Gravel' throughout Fenland. It is not yet clear how the relatively high sea-level required by this explanation could occur during a cold phase. However, sea-level is related to the total volume of water locked up in global ice at any given time, and the rapid onset of cold conditions, or a regional event such as the diversion of the Gulf Stream, could cause marked differences in the rates of growth of mid latitude and polar ice sheets and give rise to an apparent local mismatch between climate and sea-level.

Extensive spreads of river gravel occur at the mouths of all the rivers that drain into Fenland and these have been collectively referred to as 'Fen Edge Gravels' in the past. The larger spreads in the present district occur at Block Fen and Langwood Fen (where the gravels have been revealed by wastage of the overlying peat), and beneath Byall Fen (where the gravels are still overlain by Recent deposits). These gravels were deposited by the Great Ouse prior to its artificial diversion around the southern side of the Isle of Ely in Romano-British times (see Chapter 7). Other large patches of river gravel occur at the mouth of the Lark at Isleham Fen (mostly still covered by peat) and Mildenhall Fen, and adjacent to the Little Ouse between Joist Fen and Brandon Creek (mostly overlain by peat). Small patches of gravel also occur along the courses of minor streams at Grunty Fen, Hale Fen, West Fen and Butcher's Hill.

A sequence of four river terraces, with the Observatory Gravels a possible fifth, was recorded in the Cambridge district (Worssam and Taylor, 1969). Only the lowest two of these are present in the Ely district where the topographical relief is much less and where the higher terraces are presumed to have been either removed by erosion or to have passed seaward into the March Gravels. The First and Second terraces of the Cambridge district are at similar levels to one another where they pass into the Ely district, and they could not be separated in the present work; their deposits have therefore been classified as combined First and Second terrace.

Terrace gravels are currently worked on a large scale at Block Fen (Plate 7) where they are up to 6 m thick and their sedimentary features are well displayed. The gravels are composed almost wholly of flint, but include small numbers of other stones including Triassic, Jurassic and Cretaceous rocks derived from the glacial deposits of the area. Trough cross-bedding is present throughout, mostly in units up to 1 m thick and a few metres in width, and suggests deposition by torrential, braided streams in a broad floodplain. Frost-wedge structures and involutions occur at several levels within the gravels and suggest that they were subjected at times to periglacial conditions. Mammalian bones have been recorded in the gravels but none was collected in the present work. The Second Terrace gravels of the Cam of the Cambridge district have yielded remains of bison, red deer, giant deer, hippopotamus, horse, mammoth and rhinoceras (Reed, 1897; Marr, 1926).

The field relationships and sedimentary features of the terraces of the Ouse and Lark in the Ely district are comparable with those described by Bell (1970) for the Ouse terraces at Earith and by Preece and Ventris (1983) from Galley Hill, St Ives (both Sheet 187), which are the upstream extensions of the gravels exposed at Block Fen; and by Sparks and West (1970) and West and others (1974) for the Wissey terraces at Wretton (Sheet 159). At Earith [388 764] the gravels contained pockets of plant-rich silt that yielded radiocarbon dates of about 42 000 BP (before present) (Bell, 1970). A study of the pollen and plant debris within these pockets enabled Bell (1970) to conclude that this detritus, mostly from dwarf willows and herbs, was washed into pools on the floodplain by spring meltwater in a climate that was characterised by warm summers and cold winters. The absence of tree pollen indicates an open, steppe environment for the floodplain and the surrounding area. Of particular interest are the presence of four species of halophyte that must have grown in damp areas with salt (Bell, 1970, p.372). At Galley Hill, Preece and Ventris (1983) recorded an extensive flora and fauna with both temperate- and cold-climate elements from lenses of organic-rich silt in the lowest part of the gravels. The fauna included 24 freshwater and 15 terrestrial species of mollusc, and bones of the vertebrates *Bos primigenius* Bojanus (Aurochs), *Cervas elephas* Linnaeus (Red deer), *Coelodonta antiquitatis* Blumenbach (Woolly rhinoceras), *Dama dama* (Linnaeus) (Fallow deer), *Mammuthus* (Mammoth), *Megaceras* (Giant deer), *Palaeoloxodon antiquus* Falconer and Cautley (Straight-tusked elephant) and *Ursus* (Bear).

At Wretton a complex sequence was recorded in the lowest terrace of the River Wissey in which periods of gravel aggradation, similar in origin and age to that described at Earith, were separated by more temperate phases when sands were deposited (West and others, 1974).

There is, therefore, clear evidence that the lower river terrace gravels of the Great Ouse and, by analogy, those of the Lark, Little Ouse and other rivers that drain into Fenland, were deposited in broad alluvial braid-plains by torrents debouching from narrow, well-defined river valleys onto the flat Fenland plain. When traced away from the valley mouths in the present district these gravel spreads appear to maintain a relatively constant maximum thickness until they reach an abrupt termination where they have thick Recent deposits banked against them. The slope of the flat upper surface of the gravels suggests that they were graded to a base level close to this surface, and their abrupt terminations suggest they were deposited as fans in standing water.

Plate 7   River Terrace Gravels at Block Fen, Mepal [431 841]. Cross-bedding visible in the lower part of the face is part of a complex pattern produced by a large number of small intersecting channels which formed a broad alluvial plain (braidplain) during cold periods in the late Pleistocene. The bedding is disturbed by frost structures (cryoturbation) in the top 1–2 m of the face (A13728)

DETAILS

March – Wimblington – Manea

Baden Powell (1934, pp.194-195) recorded up to 3 m of shelly March Gravels in pits at Town End, March [415 958 and 416 952], Wimblington [414 936], Wimblington Common [435 908 to 443 910], Brown's Hill [437 905 to 436 899], Honey Bridge [436 895], Honey Hill [438 885], Holly House Farm [431 871], and Manea [481 920 and 489 911]. None of these sections is now exposed although parts of the sequence are visible from time to time in deep drains adjacent to Wimblington Common and Honey Bridge. Shelly gravels have been collected from the Sixteen Foot Drain at and adjacent to Honey Bridge, and from Vermuyden's Drain and a gravel pit at Horseway [426 871], but these sections are now overgrown.

A borehole at the Auction Ground, March [4164 9525], proved 5.6 m of orange-brown fine to medium gravel interbedded with fine- and medium-grained sand with broken and whole bivalves common throughout, including *Cardium*, *Tellina* and *Ostrea*.

Block Fen

The River Terrace gravels of the Great Ouse have been extensively worked, and continue to be worked, in the Block Fen area beneath a thin cover of peat and in areas where the peat has wasted. Because of the proximity of the water table to the ground level the pits rapidly become flooded when pumping ceases, and are used for recreational purposes. Flooded pits occur near Hiam's Farm [421 830 and 425 830] and Block Fen Farm [430 836 and 434 841]. This last pit

has only recently been abandoned: it was worked dry by means of draglines, the water table being temporarily lowered by pumps. The current working [430 842] at Block Fen is adjacent and to the west of the flooded workings. It is also worked dry and will eventually be backfilled and reclaimed for agriculture.

The current workings at Block Fen expose up to 5 m of gravel resting on a slightly irregular surface of Ampthill Clay. The gravels thin westwards to about 3 m over a distance of about 250 m. Sections in the flooded pits to the east formerly showed up to 6 m of gravel and this is probably the maximum thickness for that area. The gravels are composed almost wholly of fine and medium gravel-sized flints with small amounts of chalk, Jurassic cementstones and oysters, quartzites of presumed Triassic origin, and rare Middle Jurassic limestones and far-travelled materials. The sedimentary features visible in the gravels vary with the positions of the working faces but trough cross-bedding, channels infilled with coarser gravel or organic-rich silt, frost wedges, and other cryoturbation structures are normally visible (Plate 7).

OTHER DEPOSITS

The river terrace gravels of the Great Ouse, Lark and Little Ouse are commonly overlain by up to 1 m of fine silty and clayey sand. Seeley (1866a) and Baden Powell (1934) described a similar deposit (brickearth) overlying the March Gravels at March. Such material is common as a surface layer on the solid and Pleistocene deposits throughout the Ely district, its composition varying in relationship to the

Plate 8   Cryoturbation features in River Terrace Gravels and gravelly Head gives rise to a distinctive 'hummocks and hollows' topography at Kennyhill, Mildenhall Fen (foreground). Many of the hollows are filled with Nordelph Peat. In the middle distance, white soils derived from Shell Marl that marks the site of the former Redmere, contrasts with the surrounding Nordelph Peat. View N from near Holmsey Green: course of Little Ouse roddon arrowed (Cambridge University Air Photo Library, BBB 52)

local underlying material. It is probably of composite origin, part soliflucted, part waterlaid and part aeolian, and was probably formed during the later cold stages of the Pleistocene (Watt and others, 1966) at times when much of Britain north of The Wash was covered by ice.

Cryoturbation structures were also formed at that time. They are well displayed as soil and crop patterns in aerial photographs (Williams, 1964) and include frost stripes and polygons on the chalk uplands and large polygons on the terrace gravels. These structures and thin late Pleistocene solifluction deposits can be seen in many of the ditches cut into the solid and the older Pleistocene deposits of the district.

Probably the most impressive late-Pleistocene features in the district are those on the partly gravel covered lower slopes of the Chalk in the Lark and Little Ouse valleys. There, clusters of low ridges and hollows (2 to 3 m in differential height) of various shapes and sizes, with up to 100 hollows per km², form an undulating topography known as 'hummocky ground' or 'hummocks and hollows' (Plate 8). Sparks and others (1972) have suggested that this relief is the result of the freezing and thawing of large masses of ground-ice that formed from ground-water percolating from the Chalk. Hummocky ground is well developed on the north bank of the Little Ouse around Clouds Farm and River Farm, where thin peat and gravelly sand overlie the Chalk. Similar features on the Chalk outcrop around Kennyhill are especially notable (Plate 8).

The final retreat of the Pleistocene ice sheets from southern Britain has been radiocarbon dated at about 16 000 years BP. At its maximum extent (about 18 000 years BP) ice

of the latest glacial episode probably extended as far south as the southern part of The Wash (Suggate and West, 1959; Straw, 1960), to within 30 km of the Ely district. As the climate ameliorated extensive thin spreads of solifluction material were probably formed throughout the district, together with the ground-ice structures and other cryoturbation features described above.

One of the unsolved mysteries of Fenland is the origin of the 'islands' that rise above the Recent deposits. These 'islands' are not the local eminences of a gently undulating topography that was partially inundated by the Flandrian transgression, but are isolated, low hills standing on an almost flat plain. This plain, cut in solid and Pleistocene deposits, appears to have been formed before the Recent deposits were laid down, and comprised a low-lying area that enabled the Flandrian transgressions to be so rapid and widespread. A feature common to parts of all the Fenland 'islands' in the Ely district is a marked change of slope beneath the Recent deposits close to the edge of their outcrop. Slopes on the sides of the 'islands' are commonly 5° to 7° whereas those on the sub-Recent surface (excluding the sides of river channels) rarely exceed 1°. This change in slope commonly occurs at about Ordnance Datum (Figure 28). Its formation cannot readily be linked with the deposition of any of the Recent deposits: neither the peats nor the salt-marsh clays that make up the great bulk of the Recent sequence have erosional bases.

In a few areas, such as parts of the eastern edge of the March-Wimblington ridge, the margin of an 'island' has been shaped by the water that deposited the First/Second Terrace gravels. However, similar gravels are absent from

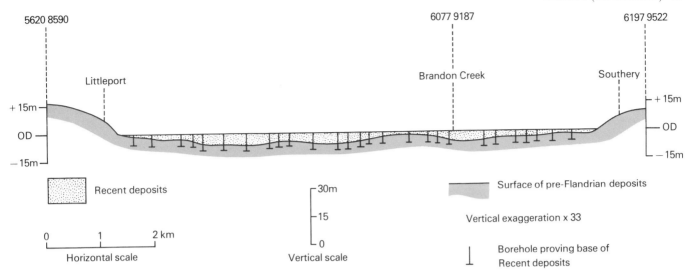

Figure 28 Section along the line of the Great Ouse between Littleport and Southery showing the relationship of the pre-Flandrian surface to the Fenland 'islands'. Based on site-investigation boreholes

other parts of the ridge and in the vicinity of most of the other 'islands'. The oldest part of the Recent sequence (see p.74) throughout most of the Ely district is a thin gravelly sand that may in part be a reworked and thinly spread remnant of what was once an extensive gravel sheet. This deposit could have been formed as river terraces or by a transgressive (prograding) storm beach moving across the area. Such a beach would produce a wave-cut platform largely free from sediment but, at the limits of the transgression, might also give rise to patches of gravel banked against the 'islands'. No gravel of this type has, however, been recorded in the district. In the absence of definite evidence it is therefore assumed that the Fenland 'islands' were formed in the latter part of the Pleistocene as true islands within a lake or shallow sea.

Faunal and floral evidence from the southern half of England suggests that the last date at which the climate was sufficiently cold for cryoturbation and related deposits to form was about 10 000 years ago (Kerney, 1963). At that time there was a rapid change to a more temperate climate and this date (10 000 years BP) has, therefore, been taken as the termination of the Pleistocene.

RECENT (FLANDRIAN)

The late Pleistocene retreat of the ice-sheets was accompanied by a corresponding rise in sea-level as water locked up in the ice was released. Sea-level rose from an estimated low of about 100 to 150 m below Ordnance Datum at the maximum extent of the ice in the late Pleistocene (Donn and others, 1962) to about 30 m below OD by the end of the Pleistocene. As sea-level rose, about 10 000 years ago, parts of the southern North Sea, including the deeper parts of what is now The Wash, became submerged and the large low-lying and low relief outcrop of Upper Jurassic clays, lying beneath what is now Fenland, became a poorly drained basin with brackish and freshwater marshes in its seaward

areas. A continued rise in sea-level caused a series of depositional zones to be established in which an open marine environment passed landward via intertidal silt and mud flats into brackish reedswamp and freshwater fens and meres (Figure 29).

In the early part of the Flandrian (10 000 to 6000 BP) a rapid and almost continuous rise of sea-level probably occurred, but this was followed by a period when rises alternated with still-stands or slight falls in sea-level. As a result, the early part of the Flandrian was characterised by rapid transgression, and the later part by alternate transgression and regression.

The history of the Recent deposits of the Ely district is, therefore, one of rhythmic alternation of marine and freshwater flooding during which the zones of sedimentation shown in Figure 29 moved to and fro across the district. As a result, marine or brackish silts and clays are interbedded with freshwater peats and shell marls (Figure 30). One major phase of regression and two of transgression can be recognised throughout the eastern part of Fenland. The oldest proven Recent deposits in the Ely district are freshwater peats that formed in river channels and poorly drained hollows about 7000 to 9000 years ago; these are overlain by an extensive freshwater peat that formed as the Fenland basin became flooded in response to the rising sea-level. The peats were inundated by a marine transgression at about 4700 BP that deposited intertidal clay and silt. Further regression and a long period of freshwater flooding between about 4000 and 2050 BP produced a second widespread peat which in turn was inundated by a major transgression that produced further intertidal deposits (Table 6).

This broad quadripartite division of the Recent deposits can be recognised throughout much of Fenland. Godwin and Clifford (1938) traced the sequence 'lower peat', 'fen clay', 'upper peat' and 'Romano-British silt' in boreholes from Littleport to Outwell, and the same sequence has been confirmed in boreholes and excavations throughout the Ely district (Seale, 1975) and in the Wisbech and King's Lynn

Figure 29 Presumed depositional
environments of the Recent deposits
of Fenland

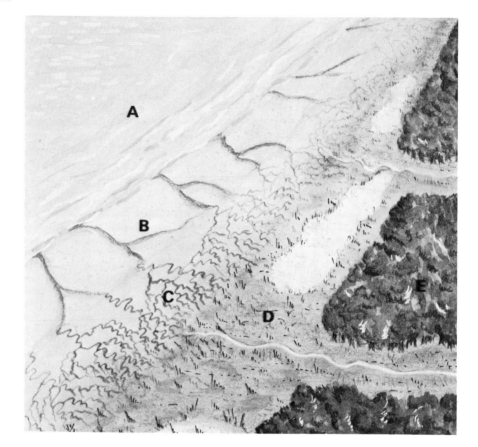

Key to depositional environments and sediment types

A Subtidal—mostly sand

B Intertidal flat—silt with clay

C Salt marsh—clay with silt

D Peat fen and meres—peat and shell marl

E Deciduous forest—erosional area

districts. The names 'Lower' Peat, Barroway Drove Beds, Nordelph Peat and Terrington Beds have been used for these four formations in the last two districts (Gallois, 1979a), and this usage is now extended to the present district. In addition, a thin gravelly sand lies beneath the 'Lower' Peat, and modern river alluvium locally caps the sequence.

Within this broad framework other minor regressions and transgressions are also present, especially around the edges of Fenland where relatively small changes in sea-level had the greatest effect on the local depositional environments, resulting in patches of peat fen or salt marsh of local extent. The size of such patches depends largely on the underlying topography. In the Ely district, where in most areas there is a gentle but steady seaward gradient beneath the Recent deposits, a few minor intercalations of peat have been recorded in the Barroway Drove Beds and of clay in the Nordelph Peat, but none of these has been sufficiently widespread to merit a separate name (Figure 31). Elsewhere in Fenland, for example in the Peterbrough district (Wyatt and others, 1984), the 'Lower' Peat and the Barroway Drove Beds have a complex interfingering relationship that gives rise to thin but extensive peats at several stratigraphical levels.

BASAL GRAVEL

The oldest Recent deposits in the district are likely to be the gravelly sands that appear to form a ubiquitous base to the sequence. These sands are known only from boreholes and the deeper drains, and are nowhere more than 1 m thick: they rest everywhere on a weathered surface of solid or Pleistocene deposits. At some localities, notably along the lines of the deeper tidal river channels, the sands are shelly and contain common *Cerastoderma*, *Ostrea* and other bivalves: they were clearly formed by tidal winnowing of the coarser debris in the channel floors. For the most part, however, the sands appear to form a thin (mostly less than 0.5 m thick) but persistent deposit, independent of topography, and seem to have been formed from local deposits as a solifluction deposit that was reworked and thinly spread at a time of considerable surface run-off. They may in part, therefore, be late Pleistocene in age and related to the formation of the Fenland 'islands' (see p.73).

'LOWER' PEAT

The oldest accurately dated Recent deposit in the Ely district is the 'Lower' Peat. As sea-level rose and the more northerly

Figure 30 Idealised rhythm for the Recent deposits showing the stratigraphical relationships of the depositional zones depicted in Figure 29

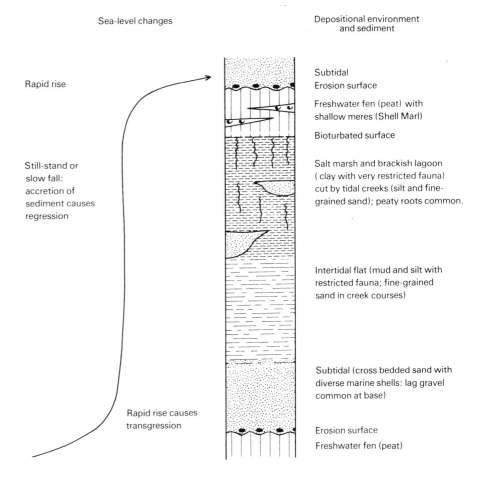

Sea-level changes

Rapid rise

Still-stand or slow fall: accretion of sediment causes regression

Rapid rise causes transgression

Depositional environment and sediment

Subtidal
Erosion surface

Freshwater fen (peat) with shallow meres (Shell Marl)

Bioturbated surface

Salt marsh and brackish lagoon (clay with very restricted fauna) cut by tidal creeks (silt and fine-grained sand); peaty roots common.

Intertidal flat (mud and silt with restricted fauna; fine-grained sand in creek courses)

Subtidal (cross bedded sand with diverse marine shells: lag gravel common at base)

Erosion surface
Freshwater fen (peat)

parts of Fenland became inundated, freshwater was ponded back in the low-lying parts of the Ely district so that the root systems of forests, which were at that time growing on the solid and the Pleistocene deposits, became waterlogged and the trees died. In many areas tree trunks are common in the basal part of the 'Lower' Peat. As the water-level rose a succession of plant communities became established, died and were preserved in the peat.

The process by which forests of oak and yew gave way to pine and finally to peat bog was graphically demonstrated in sections through the 'Lower' Peat at Wood Fen, Ely [545 851], by Skertchly (1877) and Miller and Skertchly (1878). There, five horizons of buried forest were recognised, comprising firstly a forest of tall oaks rooted in the Kimmeridge Clay, followed by smaller oak, pine and yew, and finally alder, sallow and willow (Figure 32). Subsequent studies of the pollen spectra (Godwin, 1940) confirmed this sequence, provided details of the overall flora, and traced the probable sequence of events from deciduous forest to *Sphagnum* bog. Comparison of this and other Fenland peat successions (including both the 'Lower' Peat and the Nordelph Peat) with modern examples, such as that preserved by the National Trust at Wicken Fen [553 704] in the Cambridge district, enabled Godwin (1938a) to show that this vegetational succession is part of a more complex sequence that can be traced, under the ideal conditions of no outside influence, from open water to raised bog (Figure 33).

At times when the water-level rose rapidly in Fenland, large stretches of shallow (up to 2 m deep) open water

(meres) were formed which were gradually filled by the accumulation of organic debris. In the early stages the vegetation was completely aquatic and dominated by stoneworts (*Chara*), milfoil (*Myriophyllum*), and pondweeds (*Elodea* and *Potamogeton*). In the almost total absence of waterborne clastic material, the sediments are largely organic muds and calcareous debris. This latter consists of the remains of plankton, bivalves and gastropods, but mainly of calcium carbonate encrustations of the stems of *Chara* (hence the name stonewort): the alkaline streams flowing off the Chalk provided the calcium carbonate. Skertchly (1877) recorded that in drought years, when the rate of evaporation was high, up to four inches (10 cm) of sediment composed of carbonate-encrusted stems could accumulate in a single summer.

As the open water gradually silted up the vegetation was replaced by reed (*Phragmites*) swamp and, in the drier areas, sedge (*Cladium*) fen. As the mat of peaty vegetation built up and became drier it was colonised firstly by small bushes, such as alder, buckthorn and sallow, and later by trees such as ash, birch and oak. In certain circumstances continued growth of the peat and the consequent rise in ground level caused a water-table to develop in the peat that was above, and independent of, the alkaline Fenland water. This enabled an acid peat, colonised by pine and birch, and later by heather, cotton grass and *Sphagnum* moss, to develop. Each of the plant communities described above co-existed in Fenland during the formation of the 'Lower' Peat and the Nordelph Peat (see below), their distribution being dependent upon local drainage conditions.

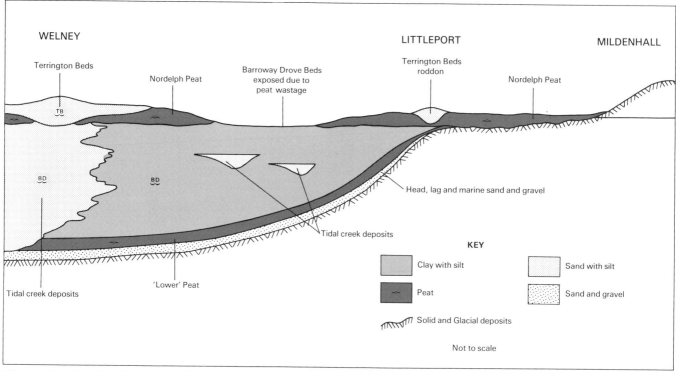

Figure 31   Section through the Recent deposits of the Welney to Mildenhall area

The 'Lower' Peat does not crop out in the Ely district, but is almost continuous beneath the Barroway Drove Beds. At a few localities near the edge of Fenland and beyond the limit of the Barroway Drove Beds transgression, the 'Lower' Peat has a narrow subcrop (up to a few hundred metres in width) beneath the Nordelph Peat. In those areas the time interval during which the 'Lower' Peat, Barroway Drove Beds, and Nordelph Peat accumulated is represented by a continuous peat succession. At a few localities where the upper levels of the peat have wasted, patches of the 'Lower' Peat may now be exposed, but they cannot be distinguished from the higher peat except by radiocarbon dating or pollen analysis. The 'Lower' Peat in such areas is likely to be represented by the trunks of large trees, mostly oak or yew, that were rooted in the underlying mineral soil; the higher peat by alder, willow, reeds or sedge.

The 'Lower' Peat is known from boreholes throughout the district and from the deeper ditches close to the edge of Fenland. It is mostly compacted and woody, and varies in thickness from a few centimetres to about 2 m, but is almost everywhere less than 1 m thick. Godwin (1940) recorded 1.95 and 1.75 m respectively in excavations in the old channel of the Little Ouse at Old Decoy Farm [660 857] and at Peacocks Farm [637 850] near Shippea Hill. Significantly, these two occurrences include two of the oldest recorded horizons of peat in the district, Pollen Zones IVa to VIIb at Old Decoy Farm and VIa to VIIb at Peacocks Farm (Godwin, 1940). They appear to have formed in the deepest part of the river channel. Elsewhere in the district, the 'Lower' Peat is mostly of Pollen Zone VIIa and/or VIIb in age. Isolated pockets of older peat (Pollen Zone VI and earlier) are preserved in the deeper river channels and in hollows in the underlying surface such as that described by Godwin

(1940) at Queen's Ground, Methwold Fens [680 930]. These older peats were formed before 7000 BP and some may be as old as 9000 BP. A radiocarbon date of about 8660 BP was recorded at Shippea Hill (Clark and Godwin, 1962).

The bulk of the 'Lower' Peat in the Ely district probably formed between about 5400 and 4700 BP. Dates of about 4840 to 5370 BP have been recorded from the 'Lower' Peat at Shippea Hill. Dates of about 4185 BP at Wood Fen, Ely and about 4535 BP at Queen Adelaide (Godwin, 1978) are from 'Lower' Peat close to the Fen edge and are near the limit of the Barroway Drove Beds transgression. This last date was from the outer rings of an enormous oak rooted into the Kimmeridge Clay. It, and similar trees, are the remnants of the forest cover that occupied the low-lying parts of the Ely district immediately prior to the inundation of Fenland.

Figure 32   Succession of forests preserved in the 'Lower' Peat at Wood Fen, Ely (after Skertchley, 1877)

BARROWAY DROVE BEDS

The formation of the 'Lower' Peat was terminated by a marine transgression which deposited the intertidal clays and silts of the Barroway Drove Beds. These beds were formerly overlain everywhere in the district by a younger peat, the Nordelph Peat, but are now exposed as small areas in Southery Fens, near Welney, and near Wimblington as a result of peat wastage. The formation lies at shallow depths beneath large areas of Southery Fens, Burnt Fen, Upwell Fen and Fodder Fen where the overlying peat is now less than 0.5 m thick. In those areas Barroway Drove Beds are exposed in even the shallowest drains.

The Barroway Drove Beds consist of soft, grey clays and silty clays which contain a complex network of channels filled with silt and fine-grained sand. Evidence from bivalves and foraminifers, and the almost ubiquitous presence of the reed *Phragmites*, indicate deposition in shallow water in salt marshes and brackish lagoons (clays) and in tidal creeks (silts and fine-grained sands). As the overlying peat has wasted and the water-table has been progressively lowered by drainage works the higher levels of the Barroway Drove Beds have become partially drained. On losing some of their moisture content they have consolidated and, because the sands and silts within the tidal creeks have lower coefficients of consolidation and compressibility than the surrounding clays, the creek deposits now form low sinuous ridges or roddons (Plate 9a). These creek patterns can be traced as pale soil marks on aerial photographs of the Barroway Drove Beds outcrop and those areas where the overlying Nordelph Peat is thin. Seale (1975), in a study of the creek patterns of the Ely district, has shown that the Barroway Drove Beds have a well defined landward limit and that a complex dendritic network of small creeks can be traced seawards. The creeks merge into ever larger creeks before passing beneath the Terrington Beds outcrop in the Welney area (Figure 34). The precise positions of the lower reaches of the larger and deeper creeks within the Barroway Drove Beds are now obscured because those same watercourses were subsequently occupied by tidal creeks during the Terrington Beds transgression. In many parts of Fenland the courses of the tidal rivers and major creeks appear to have been more or less constant in position throughout the Flandrian.

Fluctuations in sea-level and/or annual rainfall caused freshwater flooding to occur from time to time during the deposition of the Barroway Drove Beds and thin peats grew out over the marshes from the edges of Fenland. Isolated thin patches of reedy peat probably also formed within the marshes in areas where sediment accreted sufficiently quickly to raise the ground level above that of most high tides.

NORDELPH PEAT

The aggradation of the Barroway Drove Beds was ended by a regression that allowed peat to spread over the whole of their outcrop and over a large part of the adjacent Fen edge. This event was presumably caused by either a sea-level fall or a still-stand; this was accompanied by a change to a wetter climate which resulted in extensive freshwater flooding and a considerable increase in the size of the Fenland basin of deposition.

ENVIRONMENT AND VEGETATION TYPE

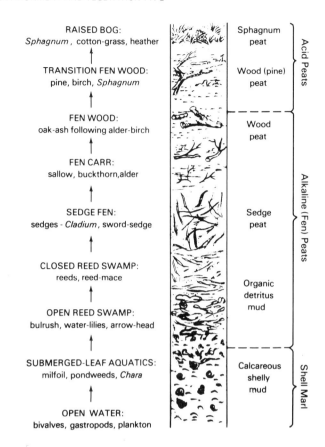

Figure 33 Natural succession of vegetation from open water to raised bog (after Godwin, 1938a)

The deposit formed by this flooding, the Nordelph Peat, originally covered about 80 per cent of the Ely district. However, wastage (mostly oxidation and ablation) since medieval times has caused large areas of peat to disappear from the Fen margin, notably from Grunty Fen, Benson's Fen and Langwood Fen (Figure 35). Much of the remaining peat is considerably reduced in thickness and is now less than 1 m thick: peat more than 2 m thick is restricted to small areas in Methwold Fens (where the land has been drained relatively recently), east of Shippea Hill (where the peat is protected by an overlying layer of Shell Marl), parts of the Bedford Washes (which have not been drained), and the deeper parts of the channels of the rivers Lark and Little Ouse.

Radiocarbon dates indicate that the formation of the Nordelph Peat began over a large area of Fenland (including the Ely district) at about 4000 BP. Over much of the district peat continued to form until it was drained in medieval times.

Freshwater flooding of the peat bog produced shallow meres from time to time, notably adjacent to the larger river roddons where drainage was impeded. Submerged-leaf aquatic plants and freshwater fauna thrived and, because the surrounding peat bogs acted as an efficient sediment trap, the meres became silted up by calcareous mud and shell debris before developing again into reed swamp (Figure 33). Where these muds are predominantly composed of shell

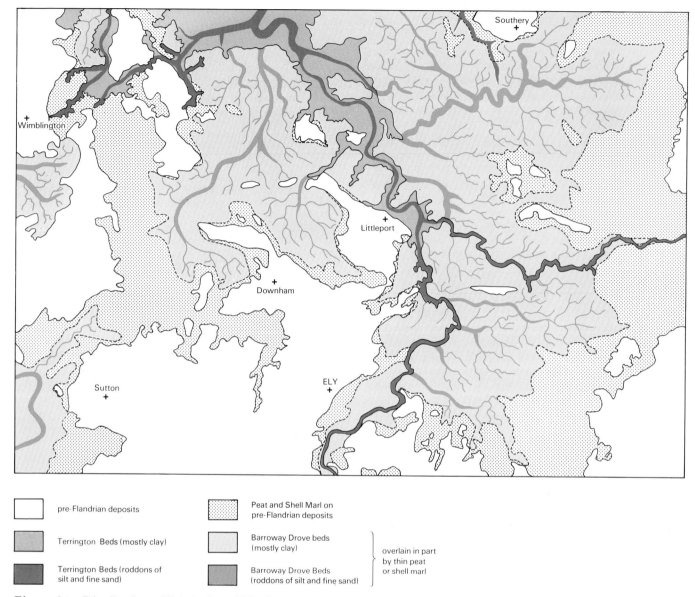

	pre-Flandrian deposits		Peat and Shell Marl on pre-Flandrian deposits
	Terrington Beds (mostly clay)		Barroway Drove beds (mostly clay)
	Terrington Beds (roddons of silt and fine sand)		Barroway Drove Beds (roddons of silt and fine sand)

overlain in part by thin peat or shell marl

Figure 34   Distribution of lithologies within the exposed Recent deposits (after Seale, 1975)

debris from *Bithinia*, *Valvata* and *Planorbis* they have been referred to as Shell Marl: where predominantly composed of encrustations of the stems of the stonewort, as *Chara* Marl. In the present account the term Shell Marl has been used to describe both deposits.

Patches of Shell Marl occur at a number of levels within the Nordelph Peat of the Ely district and range in size from a few metres to more than 4 km across. Only the larger outcrops are shown on the geological map. By far the largest is that adjacent to the Little Ouse in the Shippea Hill Mill-Bedford Farm area. This is the site of the medieval Redmere or Reed Mere (Fowler, 1947), the largest of the former Fenland meres and reputedly once the second largest freshwater lake in England (Plate 8). It was probably formed by water ponded against the roddon formed by the silts that infill the former course of the Little Ouse (Jennings, 1950). A smaller outcrop of Shell Marl near Temple Farm, also adjacent to the Little Ouse, probably marks the site of Lesser Redmere. Other patches of Shell Marl occur adjacent to the present day Great Ouse near Southery and the former course of the Great Ouse in South Fen.

TERRINGTON BEDS

In the north-western part of the district, around Welney, the formation of the Nordelph Peat was terminated by the second major transgression of the Recent sequence. This transgression converted the freshwater fen once more into a salt-marsh drained by tidal creeks. The major rivers, which had probably remained as open waterways during the deposition of the Nordelph Peat, became tidal throughout the district. In the Welney and adjacent areas, and in the river channels and tidal creeks, the Terrington Beds were deposited as silty clays (salt-marsh) and silts and fine-grained sands (tidal creeks and rivers). In that part of Fenland which lay beyond the southern limit of the transgression, the Nordelph Peat continued to form.

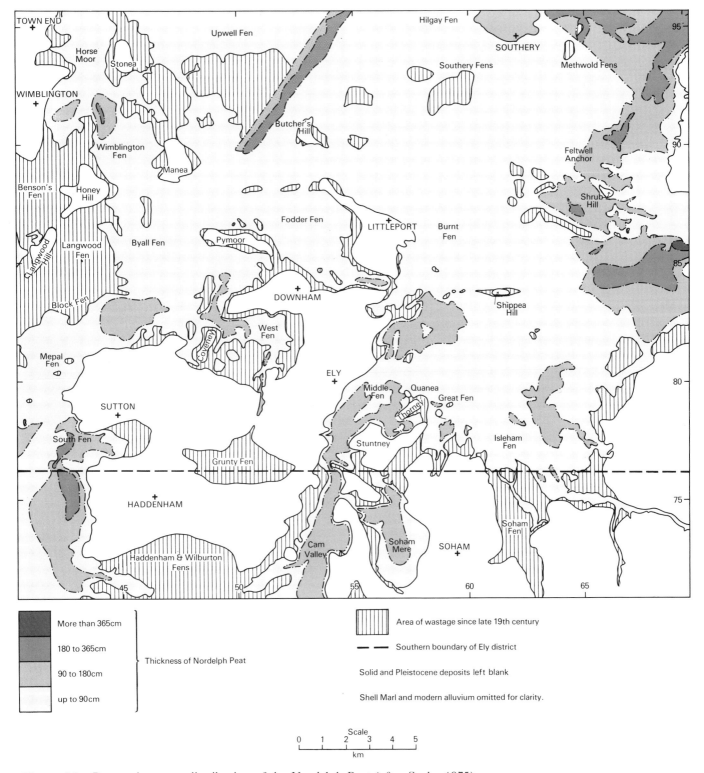

Figure 35 Past and present distribution of the Nordelph Peat (after Seale, 1975)

The main Terrington Beds transgression has been radiocarbon dated at between about 2050 BP (Willis, 1961) and 2600 BP (Godwin, 1978). The deposits that it produced were sufficiently well drained by the time of the Roman occupation for them to have been extensively colonised, but a second lesser transgression in the Third Century deposited silts and clays over many of the Roman settlement sites and filled in many of their drainage works.

As with the Barroway Drove Beds, drainage and consolidation of the Terrington Beds has caused an inverted relief to develop so that the former courses of the rivers and creeks are marked by roddons (Plate 9b). This effect has been greatly exaggerated in the case of the Terrington Beds by shrinkage and wastage of the underlying Nordelph Peat, with the result that some of the former river channels are now more the 4 m higher than the surrounding land. Fowler

Plate 9a Wastage of the Nordelph Peat has revealed a complex pattern of silt-filled tidal creeks within the underlying Barroway Drove Beds. As the peat continues to waste and the Barroway Drove Beds clays that enclose the silts consolidate, the silts are beginning to form low-relief roddons. View E across Methwold Fens from near Brandon Creek (Cambridge University Air Photo Library, AXI 46)

Plate 9b Junction of the Great Ouse and Little Ouse roddons near Sandhill, Littleport. The sinuous courses of the roddons are revealed by the contrast between their pale coloured silt infillings and the surrounding black peats. The dark stripe in the centre of the Little Ouse roddon marks the position of the last open channel, now infilled with peat. To avoid flooding, farm buildings are concentrated on the higher ground of the roddon. View ESE along the Littleport to Shippea Hill Road (A1101) from Littleport Bridge (Photo: Cambridge Air Photo library, AXI 51)

(1932; 1934b) believed that the present-day shape of the roddons could be explained entirely by peat wastage (Figure 36). However, Godwin (1938b) presumed Fowler's explanation to be wrong because it relied wholly on peat shrinkage even though in some sections across roddons the shape of the roddon demonstrably does not equate with the thickness of the underlying peat (Figure 36). Godwin suggested that the roddons were the natural levées of tidal rivers that had always stood above the level of the surrounding peat fens through which they ran. Neither explanation is wholly satisfactory because no account was taken of the differential consolidation or compaction that must have occurred between the salt-marsh clays and the silts in the tidal channels. Consolidation of the upper layers of the Barroway Drove Beds (in which there is no peat to either shrink or waste) has produced a network of roddons, albeit mostly less than 1 m high, that many local farmers maintain were absent prior to the deepening of the drainage ditches after the Second World War.

Small parts of the present differences in heights of the roddons and the surrounding peats may be due to original differences in the levels of the creek/river silts and the surrounding peats due to natural levées, but by far the greater part of the differences must be due to a combination of peat shrinkage/wastage and the differential compaction/consolidation of the Barroway Drove Beds and Terrington Beds.

Terrington Beds silts and sands, deposited at the time of the maximum transgression, are preserved at heights of up to 3.5 m above OD in the former tidal river channels of the Great Ouse and Little Ouse where they cross the Ely district (Plate 9b). Even at times of heavy run-off the present-day Fenland rivers carry only clay and fine silt: the sandy sediments now being deposited in their tidal reaches are brought from The Wash by flood tides. A similar process presumably operated during the Terrington Beds transgression. The present-day rivers are confined by artificial banks and the former tidal rivers and creeks were presumably similarly confined, but by natural features, to produce the well defined distribution of the roddon silts and sands. Godwin (1938b) suggested that the rivers ran between natural levées: an alternative possibility is that they ran through peat fens that stood at about high-water mark (about 4.5 m above OD) and impeded the flow of salt-water and sediment when

Figure 36 Origin of roddons
A. Origin of Terrington Beds roddons assuming peat shrinkage to be the major factor (after Fowler, 1934a)
B. Section across a roddon in the Barroway Drove Beds (after Seale, 1975)

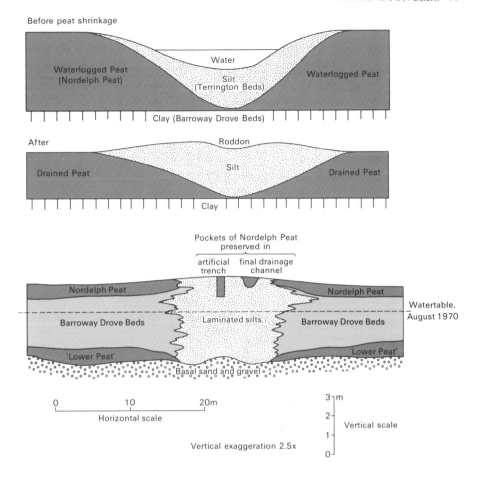

Plate 10 The vegetation on the modern Alluvium deposited by floods between the Old and New Bedford rivers contrasts with the black peaty soils of the adjacent South Fen. View NE towards Sutton (Cambridge University Air Photo Library, BPK 50)

the tide was in flood. In the medieval period, when parts of the peat fen were drained for the first time, many of the former tidal rivers and creeks were still sufficiently free of sediment to form the basis of drainage works. The present day position of these channels, on the crests of roddons (commonly visible as dark, peat-filled features on aerial photographs), would make such usage impossible: it must be assumed, therefore, that there was formerly a sufficient fall in level (presumably with the aid of simple sluices) for the surrounding peat fen to drain into them.

ALLUVIUM

The youngest deposit in the Ely district is freshwater Alluvium deposited by floodwater during the last 300 or so years between the banks of the Old and New Bedford rivers between Mepal and Welney (the 'Ouse Washes') and between the confining banks of the Great Ouse at and south of Ely (Plate 10). Seale (1975, pp.24–25) recorded up to 0.6 m of silty clay in the 'Washes' near Mepal and estimated the maximum rate of accretion at about 2 mm a year.

DETAILS

Exposures in the Recent deposits are almost wholly confined to drainage ditches; because of the high water-table and the rapidity with which they become weathered and vegetated, individual sections commonly degenerate from perfect exposure to obscurity within a year. Most of the sections described within the uppermost 2 to 3 m of the Recent deposits of the district (e.g. those in Skertchly, 1877; Godwin, 1978; Seale, 1975) have been of this type. Archaeological excavations (e.g. Clark and others, 1935), hand auger borings (Seale, 1975) and site- investigation boreholes have been used in conjunction with the ditch exposures to determine the stratigraphical sequences within the Flandrian, and to plot their regional distribution (Figures 34, 35 and 36). Summaries of the borehole sequences are given in Appendix A.   RWG

CHAPTER 7

Economic geology and land use

The Ely district contains within it the southern and eastern limits of Fenland. It attracted successive waves of settlers who used its complex network of waterways partly as trade routes and partly for defence. Agricultural settlements were probably established on the higher ground at Ely and on most of the Fenland 'islands' in Neolithic times. The Romans extended these settlements and improved the drainage and navigation systems, but much of their work was inundated during the Dark Ages due to a combination of neglect and rising sea-level.

In the 8th to 10th centuries the higher parts of the district were invaded by the Saxons, Vikings and Danes and much of the present-day pattern of settlement was established. Drainage and reclamation works begun by the early settlers have been carried on, with periods of activity separated by periods of inactivity and deterioration of the works, until the present day.

The Ely district is almost exclusively agricultural, has a low population density and virtually no mineral industry. Not surprisingly, therefore, the present-day economic geology and land use still reflect those of the early medieval period. In the past, gravel, chalk, brick clay, sand, building stone and phosphate have been worked on a small scale, and peat was once extensively dug for fuel. Small amounts of gravel continue to be worked for aggregate. The water supply of the district was formerly obtained from shallow wells but is now piped into the area from boreholes in the Lower Chalk of adjacent districts.   RWG

LAND USE

Settlement

Little is known about the early settlement of the Ely district, largely because most of the land available during the Palaeolithic and Mesolithic periods has been covered by younger Recent deposits. Scattered Mesolithic and Neolithic finds have been recorded from the 'Lower' Peat and from raised gravelly areas at a few localities (Clark and Godwin, 1962), notably from low ridges of terrace gravel in the Shippea Hill area and in excavations at Peacock's Farm, Shippea Hill [629 847] (Clark and others, 1935).

By contrast, Bronze Age remains are relatively common around the south eastern edge of Fenland and for some distance into what were then the peat fens (Fox, 1923). This distribution has been taken to indicate a warm, dry climate that allowed much of the fen to become accessible. The Nordelph Peat has yielded Bronze Age flint implements at Shippea Hill (Clark and others, 1935), a collection of bronze tools and weapons at Stuntney (Clark, 1933) and the skeleton of a woman complete with necklace and ornamental bronze pin in Southery Fens (Fowler and others, 1931). An occupation site has been recorded on the edge of the Fen

near Mildenhall, and scattered Bronze Age finds have been noted at Manea, Wimblington and Thorney. Iron Age remains are rare in the district, and it has been generally assumed that a change to a wetter climate and/or a more rapid rise in sea-level caused the peat areas to become waterlogged and largely inaccessible (Table 6). Iron Age settlements appear to have been restricted to the Fenland 'islands' and the uplands: they have been recorded at Honey Hill, Manea, on the March-Wimblington ridge and at Stonea. This last named locality includes the Iron Age fort known as Stonea Camp.

The Romans were the first to attempt systematically to drain parts of the peat fen to farm its rich soils. The present day landscape and land use of the Fenland part of the district is the result of a history of reclamation that was begun early in the Romano-British period and still continues. In scale and complexity it is unmatched anywhere in England except in some of the adjacent Fenland districts. It is not yet clear whether the flooding that caused the Iron Age population to retreat from Fenland continued into the Romano-British period, but it seems reasonable to assume that, during the early part of the Roman occupation, conditions were drier than they had been and large areas were again accessible. Nevertheless, the Romans had to undertake extensive drainage works to make farming possible. In addition to numerous minor drains and field ditches, these works probably included the diversion of the Great Ouse from an outfall into Fenland at Earith to a new course around the south side of the Isle of Ely to join the Cam at Stretham, and the diversion of the Lark from its natural channel into a canal. Other probable, but as yet undated, works of Roman origin include the diversion of part of the Little Ouse and the construction of a link between the Great Ouse at Littleport and a tidal creek (that gave access to The Wash at what is now King's Lynn) at Brandon Creek (Figure 37).

A large number of Roman settlements have been recorded in the Ely district, particularly on the Terrington Beds outcrop along the line of the former course of the Great Ouse between Littleport and Welney, where a number of sea-salt works were situated. Evidence of an extensive agricultural industry occurs on both the Nordelph Peat and the Terrington Beds, and settlements occur on most types of deposit in the district. A comprehensive account of the Roman settlement of Fenland, including detailed maps showing the locations of the known drainage works, field patterns, habitation sites and industrial sites is given in Phillips (1970).

Little documentary evidence has survived concerning the medieval settlement of the district. Renewed flooding of Fenland in the latter part of the Romano-British period appears to have destroyed (mostly by silting) much of the Roman drainage work, and settlement once more became concentrated on the higher ground. The pattern of modern settlement was probably formed at about that time and later

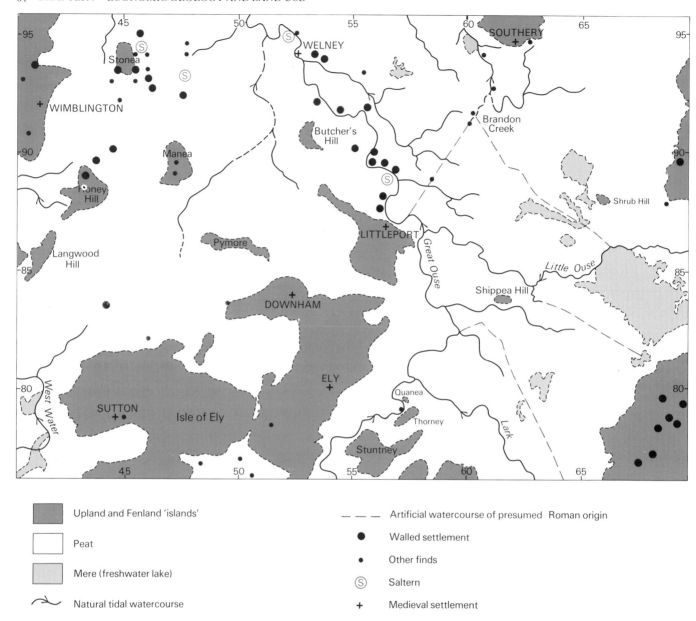

Figure 37 Geography and settlement in Romano-British times (based largely on Seale, 1975 and Phillips, 1970)

reinforced by successive waves of Saxon and Viking invaders, with the result that Ely and all the major villages of the district were established by the time of the Domesday survey (Darby, 1971, fig. 79).

The earliest published maps of Fenland, those of Hondius in 1632 and Blaeu in 1648, which were probably based on a survey of about 1604 by Hayward, show an extensive network of drains, but also large areas of undrained peat fen and other parts subject to winter flooding. The drainage of the peat fen began in earnest in the 17th century under the direction of the Dutch engineer Vermuyden with the construction of the Old and New Bedford rivers, Vermuyden's Drain, Sixteen Foot Drain, Sam's Cut, Grunty Fen Drain, and improvements to the existing drains. These works still form the basis for the modern drainage of the district (Figure 38).

Contemporary accounts of the 17th century works have been given by Vermuyden himself (1642), Dugdale (1662) and Dodson (1665). A detailed account of the history of the drainage of Fenland, including a discussion of the field and documentary evidence relating to the positions of the older drains and river diversions, has been given by Skertchly (1877). Numerous later accounts have dealt with particular aspects such as natural waterways (notably Fowler, 1932, 1933, 1934a and b), drainage (Wells, 1830; Darby, 1940a), peat wastage (Hutchinson, 1980), and life in the peat fen (e.g., Astbury, 1958; Barrett, 1963, 1964; Darby, 1940b; Godwin, 1978; and Wentworth-Day, 1954). A comprehensive review of the drainage history of the Ely district and the consequent development of its agriculture has been given by Seale (1975, pp.37–68).

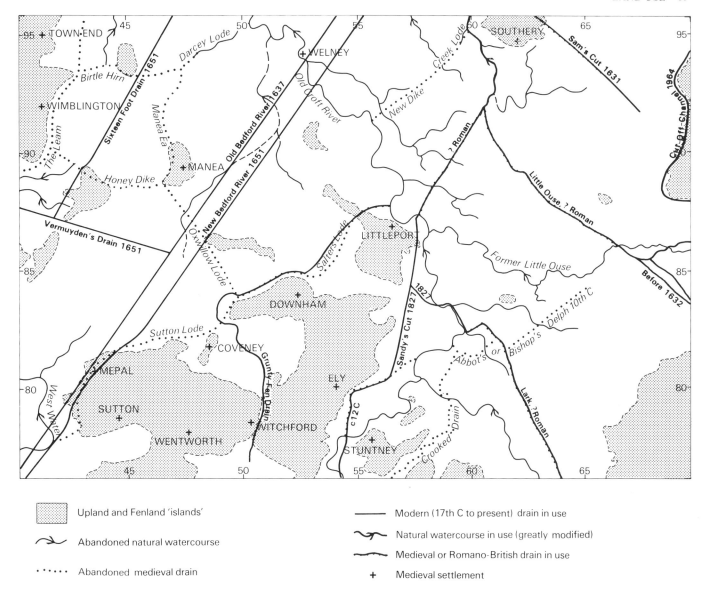

Figure 38   Principal drainage features of the Ely district, past and present (based largely on Seale, 1975)

Soils

Soils are produced by a combination of processes, of which the two most important in the Ely district are the chemical and mechanical weathering of the geological formations, including the redistribution of organic matter. The most important weathering process in the district at the present time is chemical weathering by groundwater containing dissolved oxygen and carbon dioxide. This has different effects on different rocks.

In the older argillaceous formations (Ampthill Clay, Kimmeridge Clay, Gault) and in the Boulder Clay that contains these rocks in its matrix, chemical weathering causes hydration, a general decrease in crystallinity of the clay minerals, and oxidation of pyrite to release acids that react with calcium carbonate in the matrix and fossils to produce selenite. As a result, the original soft shelly grey mudstones are converted to yellow and brown sticky clays. The younger

argillaceous formations such as the Barroway Drove Beds occupy low-lying ground with high moisture and organic contents and are less susceptible to oxidation. The sandy formations (Sandringham Sands, Woburn Sands and Carstone) contain large quantities of iron minerals such as glauconite and limonite and lesser amounts of siderite and chamosite. These minerals become oxidised to form finely divided hydrated iron compounds that are commonly leached away in solution or as colloids, leaving a yellow or brown sand residue. The calcareous formations (Chalk, and the more calcareous parts of the Upper Jurassic clays and Gault) are slowly dissolved and any included clastic or mineral matter accumulates as a surficial residue.

The establishment of vegetation intensifies the rate of weathering and produces a pedological soil. Successive generations of plants and animals add humus to the soil, and cause a concentration of some minerals and depletion of others. The long term effect of these changes is to produce a soil profile, a succession of roughly parallel layers, in which successive changes from unweathered rock to surface soil can be traced.

Within the present district variations in soil type are related to two main factors, the geology and the surface relief. A total of 41 types of soils were recognised in the district by the Soil Survey of England and Wales, and the classification and detailed description of each soil is given in their memoir for the Ely district (Seale, 1975). The main soil series are described in Table 7 and their distribution is shown in Figure 39. Four types of soil dominate. Two of these, organic soils and gley soils, cover most of the area. The remaining two, brown earths and calcareous soils, are much less extensive. The four are described below.

Three main types of site occur within the Ely district:
(i)   Well drained sites such as those on the Woburn Sands, Chalk and Glacial Sand and Gravel, where the water-table is permanently at some depth below the soil profile;
(ii)   Sites on clays such as the Upper Jurassic clays, where perched water-tables occur seasonally because of the impermeable nature of the subsoil.
(iii)   Low-lying areas, where the water-table is within or close to the soil profile for most of the year.
Site-types (i) and (ii) are restricted to the Upland, the Fenland 'islands' and the Chalk escarpment of the eastern part of the district, and type (iii) to Fenland. An intermediate type of landscape, the 'Skirtland', occurs as poorly drained, low-relief areas both around and within Fenland where the Nordelph Peat has wasted away and revealed the underlying deposits.

Organic Soils form on peat (Nordelph Peat) and cover about 55 per cent of the district. These soils are the richest in the district and produce a wide range of crops, mainly potatoes, sugar beet and cereals, but including significant

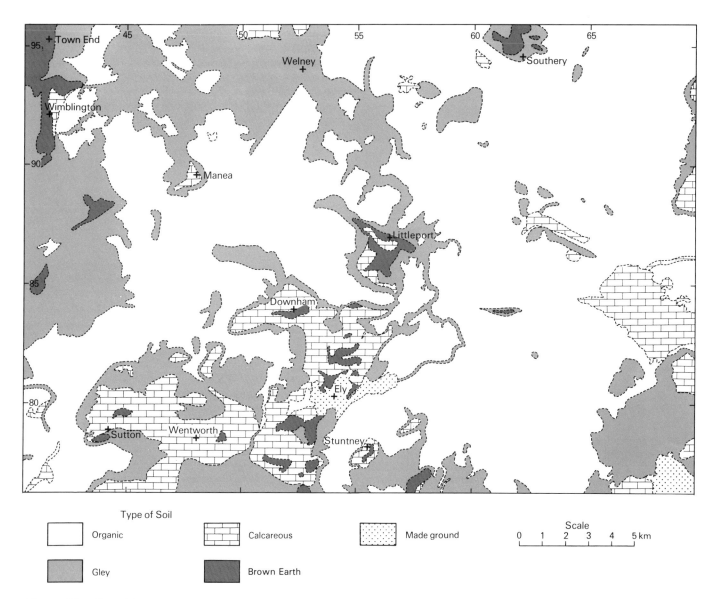

Figure 39   Simplified sketch map of the soils of the Ely district (based on Seale, 1975)

amounts of onions, peas, carrots and celery. Because of their value as arable land only small numbers of livestock are kept on them, mostly cattle and sheep to graze the banks of the numerous drains. The areal extent of the organic soils of the district is slowly diminishing as the peat wastes due to oxidation, bacterial action and ablation (see p.88 for details), and the underlying formations are gradually revealed. Where the latter come within reach of the plough the peat becomes mixed with the subsoil so that a humus soil is retained for some years after the peat has disappeared. In many skirtland areas, such as Grunty Fen, the soils now contain little, if any, more organic matter than the corresponding upland soils. It has been estimated that 120 000 acres of the 330 000 acres of peat that were present in Fenland at the beginning of the century have been converted to skirtland (Mason *in* Seale, 1975, p.67).

Gley Soils on both upland and fen are the second most extensive group in the district and cover about 30 per cent of it. They are gradually replacing the organic soils as the Nordelph Peat wastes and will eventually occupy about 85 per cent of the area of the district if the peat is not protected. Gley soils result from the reducing action of bacteria, micro-organisms and plant roots acting upon ferric and other compounds in poorly drained conditions. Such soils are commonly grey but may contain brown limonitic concentrations in their upper layers if intermittent waterlogging produces alternately reducing and oxidising conditions. The Upper Jurassic clays and the Gault give rise to gley soils (surface-water gley soils) because of their impermeable nature.

Downward percolation is restricted with the result that a temporary water-table is present in winter a little below ground level; in summer, as the clays dry out, lateral drainage occurs through the cracks formed in the surface layers. All the formations, except the peat, Shell Marl and a small part of the Terrington Beds, that crop out in Fenland, including those in the skirtland area from which the Nordelph Peat has wasted, give rise to gley soils (ground-water gley soils and humic gley soils) because of the permanent proximity of the water-table to ground level. All the skirtland soils formerly had permanent water-tables at or above the present ground level and are now pump-drained to enable them to be cultivated. The gley soils of the Fenland 'islands' support predominantly arable cultivation, with cereals as the most important crops. Potatoes and sugar beet have recently declined and beans have been largely replaced by oilseed rape as a catch-crop. The farming there is more mixed than on the organic soils, and grassland and small herds of cattle are relatively common. In the skirtland areas the crops grown are comparable to those of the adjacent islands. The more silty parts of the Terrington Beds outcrop give rise to a rich silt soil that is suitable for a wide range of vegetable crops and for selected fruits such as strawberries.

Calcareous Soils form on the more calcareous parts of the Ampthill Clay, Kimmeridge Clay and Gault, and on the Lower Chalk, chalky Boulder Clay and Shell Marl where these are relatively well-drained and not overlain by non-calcareous superficial deposits. Such soils occupy about 12 per cent of the district. In exceptionally well-drained areas

Table 7 Simplified classification of the soils of the Ely district showing the relationship of soil type to geology and land-scape

Landscape	Parent material	Soil group	Soil subgroup	Area % of district
Upland	Lower Chalk, chalky drift	Calcareous soils	Rendzinas	13
			Brown calcareous soils	
Fenland	Shell Marl		Gleyed rendzinas	
Upland and Fenland 'islands'	Boulder Clay; Upper Jurassic clays and Gault where calcareous		Gleyed brown calcareous soils	
Fenland	Terrington Beds (where calcareous)			
Upland and Fenland 'islands'	Woburn Sands, Roxham Beds, sandy drifts	Brown earths	Normal brown earths	3
	Sandy drift on Boulder Clay or Terrace gravels		Gleyed brown earths	
	Jurassic clays and Gault	Gley soils	Surface-water gley soils	30
Fenland	Alluvium, Terrington Beds, Barroway Drove Beds		Groundwater gley soils	
Skirtland	Clayey, sandy, silty and calcareous drifts overlying various solid deposits		Humic gley soils	
Fenland	Peat	Organic soils		54

on the steeper chalk slopes, thin soil profiles are present (rendzinas) in which an organic/mineral-rich surface layer rests on chemically unweathered rock with no intervening weathered layer or with only a poorly developed one. On the lower slopes and on the skirtland, where a seasonal water-table is present, gleyed rendzinas are formed on the Chalk. On well-drained sites where the calcareous content of the Chalk or Boulder Clay is mixed with sand and stones from local drift deposits, a more complete soil profile is present and brown calcareous soils are formed. The same rocks give rise to gleyed brown calcareous soils in imperfectly drained or low-lying areas.

Brown Earths are well-drained soils that form on the Roxham Beds, Woburn Sands, and the sandy non-calcareous drift deposits; their imperfectly drained equivalents, the gleyed brown earths, form on sandy and gravelly deposits (mostly river terrace gravels) in the skirtlands. Together, they occupy about 3 per cent of the district. The calcareous soils and brown earths of the district support predominantly arable, mixed farming which includes sugar beet and potatoes as well as cereals.

Two inter-related problems of a geological nature continue to affect farming in the Fenland part of the district: both result from the interruption to the natural processes of sedimentation caused by draining the area. Firstly, peat wastage is causing some of the richest soils of the district to be replaced by much poorer soils. Secondly, peat shrinkage, and to a lesser extent consolidation of the underlying Barroway Drove Beds clays, is lowering the ground level in many areas, making drainage more difficult and flooding more likely.

The agricultural problems that can arise from ablation of the peat by strong winds in the Spring (the 'Fen Blow') have been summarised by Mason (*in* Seale, 1975) as loss of small seeds such as carrots, onions and sugar beet (during severe 'blows' larger seeds such as cereals and peas may also be lost), damage to seedlings either by the wind itself or by burial beneath blown soil, loss of topsoil (including fertilizers), and infilling of ditches by soil and peat. In the past, clay was commonly mixed with the peat to stabilise it, but this method, although effective, has now become prohibitively expensive. Netting is sometimes used as a windbreak to protect small areas of valuable crop and, on a larger scale, a fast growing crop such as mustard is sown in rows among the crop to be protected so that it can act as a temporary barrier during the period when blows are likely to occur. In the absence of windbreaks the most effective preventive measure to avoid wind erosion is to prepare and seed the soil in the shortest possible time so that the surface layers suffer the least physical disturbance and retain as much of their moisture content as possible.

In many areas the moisture content of the Nordelph Peat is more than 100 wt per cent and that of the Barroway Drove Beds more than 40 wt per cent. When drained, both shrink by amounts proportional to their loss of moisture content and, in the case of a thick bed of peat, this can cause a fall in ground level of more than 1 m. When the peat-fen formed about 500 to 2500 years ago its water-level was up to 4.5 m above OD, but centuries of drainage, combined with wastage, have reduced ground levels over large areas of the Ely district to below sea-level (Figure 1). If the Nordelph

Peat eventually wastes away completely, large areas of solid and Pleistocene rocks will become exposed in the district (Figure 40).   RSS, RWG

CONSTRUCTION MATERIALS

Building stone

There is no rock at outcrop within the Ely district that is sufficiently strong to be used on its own as building stone. In consequence, all the more imposing older buildings in the district, notably the Cathedral and Bishop's Palace at Ely and the churches at Downham, Littleport, March (Town End), Southery, Sutton, Wimblington, Witham and Witchford, are constructed of a framework of imported stone (mostly Lincolnshire Limestone from the Peterborough area) with local materials used to infill the walls. The humbler and more recent buildings are made of brick (commonly with a timber frame in the older examples) or of a combination of brick and local stone.

The most commonly used local material is the Woburn Sands. Where unweathered, the sands are grey or greenish grey, and calcareously cemented to form a weak sandstone that can be easily worked into roughly dressed blocks that weather to a pale yellowish brown on exposure. The cathedral at Ely is founded on this material, and large quantities of it, presumably dug from the immediately adjacent area, have been used as wall-filling in the cathedral, almshouses and bishop's palace. The main fabric of the cathedral is made of Barnack Stone, a fine-grained oolitic limestone from the Middle Jurassic Lincolnshire Limestone, that was quarried on the banks of the River Welland near Stamford and transported to Ely by barge through the Fenland waterways. Other imported stones used in the cathedral are varieties of Lincolnshire Limestone (mostly used for repairs and additions after the 15th century when the 'Barnack Stone Quarries' became exhausted), including Clipsham Stone, Ketton Stone, Casterton Stone, Ancaster Stone and probably Lincoln Stone. Burwell Rock (a hard bed within the Totternhoe Stone) from the Lower Chalk of the Cambridge district, Purbeck Marble (bivalve and/or gastropod-rich limestones from the Upper Jurassic of the Swanage area), Mansfield Stone (Permian dolomitic sandstone) and York Stone (fissile Coal Measures sandstone) have been used internally.

Roughly dressed blocks of calcareous sandstone from the Woburn Sands predominate in the wall-fillings of most of the churches in the district. The composition of the remaining material reflects to some extent the availability of hard materials in the particular area. For example, Upper Jurassic cementstone and ferruginous pan (both common in the local Boulder Clay) are common in Littleport Church; flints, well rounded quartzites and vein-quartz (from the Glacial Sand and Gravel) are common in Downham Church; and flint, brashy Jurassic limestone and calcreted chalk (from the local Boulder Clay) are common in Sutton Church.

Bricks etc.

Bricks, and probably tiles and drain-pipes, have been made from the Ampthill and Kimmeridge clays at a number of

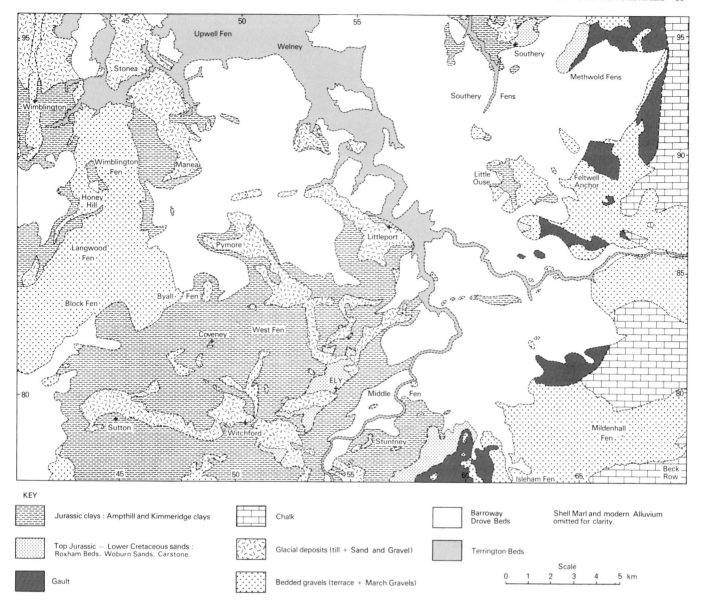

KEY

Jurassic clays : Ampthill and Kimmeridge clays	
Top Jurassic — Lower Cretaceous sands : Roxham Beds. Woburn Sands. Carstone.	
Gault	
Chalk	
Glacial deposits (till + Sand and Gravel)	
Bedded gravels (terrace + March Gravels)	
Barroway Drove Beds	Shell Marl and modern Alluvium omitted for clarity.
Terrington Beds	

Scale
0   1   2   3   4   5 km

Figure 40 Geological sketch map of the Ely district assuming complete wastage of the Nordelph Peat (based on Seale, 1974 and 1975)

localities in the district, mostly before the First World War. In the Kimmeridge Clay the more calcareous parts of the sequence (notably Beds KC 18, 30 and 44) appear to have been favoured. This is probably due to the fact that much of the Kimmeridge Clay can be satisfactorily shaped into ceramic products only at relatively high moisture contents, but the loss of moisture on firing then causes much shrinkage and cracking. The more calcareous beds generally require lower firing temperatures due to the fluxing action of their contained calcium carbonate and are, therefore, less susceptible to shrinkage. They give rise to a yellow or brown stock brick.

The Ampthill Clay was worked for brickmaking at Manea [482 891], and the Kimmeridge Clay at Haddenham End [464 766], Ely [557 816], Downham [531 845], Chettisham [565 840], Littleport [564 849 and 567 854] and Southery [616 958].

Embanking materials

Where impermeable materials have been required for repairing the banks of the Fenland drains, borrow-pits have been dug in the nearest accessible clay formation. The Ampthill Clay has been dug at Toll Farm [435 814] and at 'Gault' Hole [437 805], Mepal, and at Bury Lane, Sutton [432 793], for repairs to the Old and New Bedford Rivers; the Kimmeridge Clay at Stretham [516 743] for the Cam; the Kimmeridge Clay at Roslyn Hole, Ely [555 808], and Black Horse Farm, Brandon Creek [608 912], for the Great Ouse; and the Gault at Castles' Farm, Soham [600 773], and Decoy Farm, Little Ouse [651 865], for the Lark and Little Ouse respectively.

Glacial Sand and Gravel was dug at Downham [528 842], as was a mixture of Boulder Clay and Kimmeridge Clay at

Chettisham (547 838] for use in embankments on the Ely to March railway where it crosses Fenland.

Aggregate, ballast and building sand

All the gravelly and sandy formations of the district have at some time been worked on a small scale to provide local construction materials for roads and buildings. A small pit at Southery [622 948] worked the Roxham Beds for building sand, and sand was probably obtained as a by-product from most of the building-stone workings in the Woburn Sands and from gravel workings. The March Gravels were worked in large pits at Wimblington Common [435 900 to 438 913] and in smaller pits at Horseway [426 871], on the Wimblington-March ridge, at Cow Common [489 911] and the Dams [478 925]. The most extensively worked deposits in the Ely district, and the only ones that continue to be worked, are the River Terrace Gravels of the Block Fen area. Large flooded gravel pits occur near Hiam's Farm [421 830 and 425 830] and Block Fen Farm [433 840]; gravel continues to be worked in an adjacent pit [430 842]. The gravels are worked dry by dragline, the water-table (close to ground-level) being temporarily lowered by pumping. The gravels are up to 6 m thick, composed predominantly of flint, and are used mostly for concreting aggregate.

The unworked reserves of gravel in the Block Fen area and beneath the peat of Byall Fen are considerable. Other extensive spreads of gravel occur where the Lark and Little Ouse and their former courses and tributaries pass beneath the Recent deposits of Fenland (Figure 40), but their quality and thickness are as yet unexplored.

Engineering properties

Although the Upper Jurassic clays, Lower Cretaceous sands and the Gault have a broad outcrop in eastern England, and have been used as the founding materials for numerous structures and earthworks, little systematic work has been done on their engineering properties. Within the Ely district there are few modern civil engineering works and the only published descriptions of properties are those of the Gault from the Ely-Ouse Water Transfer Scheme (Samuels, 1975). There is a much larger amount of unpublished site-investigation data available for the Upper Jurassic and Lower Cretaceous rocks of eastern England, but much of this is of little value for detailed analysis because the tested samples commonly lack precise stratigraphical information or have been considerably modified during sampling. In many cases it is now impossible to determine whether the tested samples have been unweathered, weathered, or even reworked as solifluctied material or incorporated into some other drift deposit.

The engineering properties of the Ampthill Clay, Kimmeridge Clay and Gault are broadly similar in that all three formations are stiff to hard, fissured, normally active clays (Activity range 0.75 to 1.25). Moisture contents mostly range from 18 to 30 per cent and, when triaxially tested, samples within this range have undrained shear strengths that vary inversely with moisture content from below 100 kN/m^2 to about 400 kN/m^2. Published analyses for the Lias (Chandler, 1972) and the Oxford Clay (Parry, 1972) show

similar properties, and it has been suggested (Gallois, 1979a) that these properties are greatly influenced by mechanical discontinuities that were opened by permafrost down to about 80 m depth during the Pleistocene glaciations. The harder bands within the Upper Jurassic clays (cementstones and oil shales) and the Gault (silty mudstones) are significantly stronger than the bulk of the clays.

The properties of the Sandringham Sands and Woburn Sands are largely unknown in the district. Both formations contain beds of strong, calcareously cemented sandstone, but because of the permeable nature of the sands and the upland position of their outcrops most of these beds have been decalcified to a loose, locally clayey, sand. The near-surface layers of the Lower Chalk are also likely to have been partially decalcified and mechanically disturbed by Pleistocene frost action throughout most of the district. At depth, the properties of the chalk are likely to be comparable to those from the nearby Thetford district where Young's moduli are in the range 10 to 50 tonnes/sq cm depending on the degree and nature of the fracturing (Ward and others, 1968).

The few test results available for the Chalky-Jurassic till of the Fenland area suggest that its natural moisture content is generally in the range 15 to 25 per cent, and its shear strength in the range stiff to hard. No data are available for the Glacial Sand and Gravel: this material is everywhere loose and is composed largely of flints, but is laterally and vertically variable in its pebble and fines contents.

The engineering properties of the Recent deposits of Fenland vary considerably both laterally and vertically, reflecting the variable nature of the lithologies and modes of deposition. The Recent deposits are normally-consolidated sediments, although local leaching and chemical effects can give rise to apparently overconsolidated properties. Almost all sections contain at least some peat and clay, and the Recent deposits are generally poor founding materials (Plate 11). The engineering properties of the Recent deposits in the vicinity of the access shafts for the Ely-Ouse Water Transfer Scheme tunnel were such that the sediments had to be frozen to enable them to be excavated safely. Few soil tests have been carried out on these deposits and the general notes below are based largely on data from a variety of site-investigation sources in adjacent districts.

The Barroway Drove Beds are typically very soft, slightly silty, or finely sandy clays riddled in their upper part with water-filled peaty root-holes that extend down from the lower surface of the Nordelph Peat. Local lenses of peat and sandy tidal creek deposits occur within them. Typical natural moisture contents generally vary from about 25 to 50 per cent, with records of over 80 per cent where large numbers of root-holes are present. Dry densities are usually in the range 1.3 to 1.6 Mg/m^3 and are similar to those of the Terrington Beds, but shear strengths are lower, C_u (quick undrained) rarely more than 20 kN/m^2, because of the organic content and high moisture content.

The Nordelph Peat and the 'Lower' Peat vary in texture from woody to fibrous depending on the environment in which they were deposited and their distance from the uplands. Dry densities and natural moisture contents in the Nordelph Peat are generally high, between 40 and 90 per cent, with records of over 500 per cent occurring in the King's Lynn district. Values for C_u vary with the texture of the

Plate 11　Peat shrinkage features at Burnt Fen, Littleport [585 887]. Uneven subsidence caused by peat shrinkage has distorted (foreground) and tilted (background) some of the older houses in the district. More modern buildings (middle distance) are built on raft foundations to avoid this problem (A13738)

peat, but are rarely more than 20 kN/m² for the fibrous peats. Soil test data for the 'Lower' Peat are rare, but its moisture content is generally lower and its shear strength higher than that of the Nordelph Peat as a result of its greater consolidation.

The Terrington Beds are typically soft, dull reddish brown, slightly silty to silty clays with the silt concentrated in laminae. Locally, the silt may be replaced by very fine-grained sand. They form a thin surface layer, and samples taken for soil testing are commonly contaminated by modern worm and plant action and have been affected by atmospheric oxidation. Natural moisture contents vary with the efficiency of the local artificial drainage system and the seasonal conditions, but are generally in the range 10 to 40 per cent. Triaxial compression test results vary considerably in response to small amounts of contaminants such as peat or gravel: C_u is usually less than 40 kN/m².

OTHER MATERIALS

Lime and marl

The Lower Chalk was formerly worked for agricultural lime or marl in small pits at Thistley Green [679 767], Kennyhill [668 796], Blackdyke Farm [690 886] and near Whitedyke Farm [693 896 and 692 890]. The last-named locality included a lime kiln. The raw chalk or burned lime was probably used on the more sandy upland soils of the district. Small,

shallow pits, also dug for marl, are common on the Chalky-Jurassic till outcrops in the March, Manea, Littleport, Downham and Sutton to Ely areas.

Phosphate

Between about 1850 and 1890 the phosphatic pebbles and nodules in the Woburn Sands, Gault and Cambridge Greensand formed the basis for an agricultural phosphate industry where these formations crop out in Cambridgeshire, Norfolk and Suffolk. Contemporary descriptions of the method of working were given by Seeley (1866b), Fisher (1873) and Penning and Jukes-Browne (1881) and the value of the deposits as a source of phosphate was reviewed by Oakley (1941). A summary of the industry in Cambridgeshire has been given by Grove (1976). The phosphatic nodules, known locally as coprolites, are mostly small (1 to 10 cm across); they were dug from deep trenches and then separated from their matrix by washing. Extensive areas were worked but, because the extracted mineral was small in volume, the land was reinstated with the spoil and little evidence of the workings has survived. The peak years of production were 1873–75 (Grove, 1976, p.30). The industry declined rapidly after this, due partly to an agricultural recession but mainly to the availability of cheap foreign sources of phosphate. A few small workings were continued until the turn of the century. Phosphatic nodules were extensively worked in the Gault in the Soham area and some small workings may have occurred in this formation in the southern part of the present district. There is no evidence that the phosphatic pebbles in the Ampthill Clay were ever worked.

Peat

Peat was extensively worked for fuel in the Ely district in Victorian times and on a small scale from at least medieval times. The word peat was apparently unknown among the agricultural population of Fenland at that time, and the terms 'turf' (unweathered peat worked for fuel in 'turbaries') and 'moor' (weathered peat unsuitable for fuel) were in general use. Skertchly (1877, pp.135–139) provided a graphic description of the method of working in a large turbary at Coveney. The peat was dug in the spring and summer months and cut into blocks that were loosely stacked to dry. A good worker could dig up to 30 tons of raw peat (dried weight 8 to 10 tons) in a 14 hr day. In most years the drying process took about three months, but in exceptionally wet years little drying occurred and the blocks were destroyed by winter frosts. The dried peat appears to have been used exclusively for domestic fires and was the favourite fuel of the inhabitants of the Isle of Ely. Elsewhere in Fenland, small amounts of peat were used as fuel for brickmaking. The calorific value of the peat was approximately one third to one half that of the domestic coal on sale at the time. The heating costs for the two fuels were similar, but the peat burned with very little smoke or soot and was, therefore, much favoured by domestic staff who would accept employment at lower rates of pay in peat-burning households because less cleaning was involved (Skertchly, 1877, p.139). Little peat working continued after the First World War. Godwin (1978, p.123) recorded workings at Swaffham Fen, in the Cambridge district, as late as 1922 and a few small workings may have been active in the Ely district at that time. The most extensive workings recorded by Skertchly (1877) in the district were in the Coveney and Manea areas; later agricultural working and peat wastage has removed all trace of them.

Oil shale

Between 1916 and about 1923 over 60 cored boreholes were drilled to examine oil shale resources in the Kimmeridge Clay in the area between Kings Lynn and Southery. Rose (1835) had recorded 'inflammable schists' in a brick-pit at Southery, and Forbes Leslie (1917a and b) had suggested that large reserves of rich oil shale were present beneath west Norfolk. Three boreholes were drilled at the most southerly limit of the supposed oil shale field, one at Decoy Farm, Southery [6485 9482] and two in Methwold Fens [678 941 and 6921 9639] (Pringle, 1923). Over 80 seams of oil shale are present in the Kimmeridge Clay of Norfolk, but the seams are thin (1 to 45 cm) and separated by barren mudstone (Gallois, 1978). These seams contain only traces of free hydrocarbons but are rich in the complex organic mineral kerogen. On retorting at 400 to 500°C individual seams are capable of yielding substantial quantities of oil (some, more than 50 gallons of oil per ton of dry shale). The oil shale seams are not evenly distributed throughout the Kimmeridge Clay but are concentrated in the *eudoxus* Zone

(Beds KC 29 and KC 32), the *elegans* Zone (Bed KC 36), the *wheatleyensis* Zone (Bed KC 42) and the *pectinatus* Zone (Beds KC 45 and KC 46). Attempts at commercial exploitation in Norfolk have failed principally because of the difficulties of working such thin seams, the cost of producing the shale-oil and its high sulphur content. The oil shales are lithologically distinctive, being brownish-grey with a brown streak. They are more resistant to weathering than any other Kimmeridge Clay lithology except cementstone and are, therefore, common as erratics in the local boulder clay. The oil shales of the *eudoxus* Zone (Bed KC 32) are exposed at Roslyn Hole, Ely [552 807] but are deeply weathered there (see p.40).

WATER SUPPLY

The presence of a great thickness of Upper Jurassic clays beneath all but the most easterly part of the Ely district has meant, in the past, that the supply of water from underground sources was limited to that from shallow boreholes. In Victorian times the town supply at Ely was obtained from numerous shallow wells into the Woburn Sands and those for March, Chatteris and Wimblington from wells into the March Gravels. Coveney, Downham, Haddenham, Littleport, Southery, Stretham, Witcham and Witchford all obtained their supplies from a combination of springs and shallow wells (Whitaker, 1921, 1922). The danger of pollution of these shallow wells, at a time when the systematic treatment and disposal of sewage was unknown, was great, and a report on the water supply of Ely in 1850 (Lee quoted *in* Whitaker, 1922) noted that 'the rain-water, slops, fluid of cesspools, and all manner of refuse, soak into the soil, and converge by gravitation towards the nearest well-pump'.

The unsavoury nature of these supplies and the consequent risk of epidemics led to the establishment of safe supplies away from the centres of population. Ely and the adjacent villages are now supplied from wells in the Lower Chalk at Isleham in the Cambridge district. The supplies for Manea, March and Wimblington are obtained from boreholes associated with springs that issue from the base of the Lower Chalk at Marham in the Swaffham district. Small amounts of water are still obtained from wells in the south-eastern part of the Ely district, mostly for agricultural needs, from the Woburn Sands and Lower Chalk.

Details of the sequences in water abstraction boreholes in the district are given in Appendix 1. The few details that are known of the yields or the water chemistry of the older wells are given by Whitaker (1921, 1922). Of particular interest are a number of shallow wells into the Quaternary deposits in the Byall Fen area, notably that near Welches Dam [4572 8612], in which Harmer (1871) recorded water at 69°F when that in the adjacent drains was at 38°F. Skertchly (1877, p.243) concluded that a deep-seated source was unlikely and favoured Fisher's (1871) theory that the heating effect was due to organic decomposition. No chemical data were produced to support this suggestion, but a near surface effect is now generally thought to be the cause.

No analytical detail is available for the waters obtained from the Woburn Sands and Lower Chalk in the district, but they are likely to be similar to those quoted for these formations in adjacent districts. Wells in the Woburn Sands show a wide range of yields, and the water can vary from soft to hard. High chloride and high iron contents are locally present. Yields from wells in the Lower Chalk vary from low to very high, depending mainly on the lithology of the chalk penetrated and its degree of fracturing. Harder bands such as the Cambridge Greensand and Totternhoe Stone give significantly higher yields than the adjacent chalks. The highest recorded yield in the region is at Isleham Pumping Station [641 729], the source of supply for Ely, where an average yield of 222.8m³/h (49 000 gal/h) was recorded during a 14-day pumping trial (Worssam and Taylor, 1969, p.133). Chalk waters are invariably hard thereabouts, most-

ly due to carbonates, and may locally contain significant quantities of chlorides or sulphates.

Although the average annual rainfall is less than 600 mm, one of the lowest in Britain, the district is currently a net exporter of water. Fenland as a whole has a catchment approximately six times as large as its own area, and for part of the year large quantities of water run to waste into The Wash. Some of the excess water that would otherwise drain into south- eastern Fenland is intercepted by the Cut-off Channel and pumped to Essex as part of the Ely-Ouse Water Transfer Scheme through a tunnel and pipe-line that begins at Blackdyke Farm [691 882]. RWG

APPENDIX 1

Boreholes

The British Geological Survey holds records of over 200 boreholes in the Ely district and the information obtained from the better documented of these has been used in the preparation of the 1:50 000 geological map. The full records of all the boreholes are held on open file at the Survey's Keyworth Office. The great majority were drilled for site-investigation purposes and few provide details of the solid geology. Summaries of the geological sections proved in the stratigraphically more useful boreholes, most of which are referred to in the text, are given below. The reference number in the BGS filing system is given for each borehole (e.g. TL 69 SW/39 — Decoy Farm Borehole, Southery).

SHEET TL 47

TL 47 NW/1 to 17 — Haddenham Internal Drainage district boreholes [418 780 to 477 776]
Drilled in 1952 for site investigation. Ground levels 1 to 2 m above OD
Shallow boreholes that penetrated up to 5.72 m of Recent deposits of peat, clay and sand.

SHEET TL 48

TL 48 NE/1 — Oxlode Pumping Station [4824 8579]
Drilled in 1960 for site investigation. Ground level 0.8 m below OD.

	Thickness m	Depth m
RECENT		
Nordelph Peat	0.48	0.48
Barroway Drove Beds	2.03	2.51
'Lower' Peat	0.23	2.74
?PLEISTOCENE		
Stony clay, silt and sand	2.85	5.59
JURASSIC		
? Ampthill Clay	0.56	6.15

TL 48 NE/3 — Welches Dam [4572 8612]
Drilled in about 1870 for water. Ground level about OD.
Recent deposits of peat, clay and sand proved to c.3.1 m.
See Whitaker 1922, pp.95 – 99 for details.

TL 48 SW/1 — Wenny Farm, Chatteris [4135 8487]
Drilled in 1975 for research. Ground level about 1.5 m above OD.

	Thickness m	Depth m
PLEISTOCENE		
Cryoturbated gravelly sand, passing down into sand and gravel	c.4.5	c.4.5
JURASSIC		
West Walton Beds	c.5.5	10.0

See Gallois, 1976, p.19 for details

SHEET TL 49

TL 49 NW/6 — Auction Ground, March [4164 9525]

	Thickness m	Depth m
PLEISTOCENE		
March Gravels	5.9	5.9
Salt-marsh clay	0.5	6.4
Sand: fine-grained, interbedded with gravel	1.1	7.5
Till: Chalky - Jurassic clay with flints etc.	2.5	10.0

See Gallois, 1976, p.18 for details.

TL 49 SW/1-Wimblington Road, March [4179 9482]
Drilled in 1975 for research. Ground level about 4.6 m above OD
Chalky-Jurassic till proved to 27.0 m.
See Gallois 1976, p. 18 for details.

TL 49 SE/1-White Gate Farm, Stonea [4506 9449]
Drilled in 1975 for research. Ground level about OD

	Thickness m	Depth m
PLEISTOCENE		
Gravelly, clayey sand interbedded with Chalky-Jurassic till	5.3	5.3
JURASSIC		
Ampthill Clay	4.7	10.0

See Gallois, 1976, p.19 for details.

TL 49 SE/2 — Stonea Farm, Stonea [4574 9394]
Drilled in about 1930 for water. Ground level 3.4 m above OD.
Chalky-Jurassic till proved to 1.8 m.
See Baden-Powell 1934, p. 213 for details.

SHEET TL 57

TL 57/NW 1 and 2 — Great Ouse River Authority Boreholes 1 to 8 [5391 7714 to 5385 7688]
Drilled in 1948 for site investigation. Ground levels 1.1 m below OD to 3.9 m above OD.
Shallow boreholes that proved up to 8.5 m of Recent deposits of peat, clay and sand.

TL 57 NW/8 to 18 — Great Ouse River Authority Boreholes 30 to 40 [5403 7730 to 5448 7922]
Drilled in 1950 for site investigation. River bed levels 1.1 to 1.5 m below OD.
Shallow boreholes (drilled in the river bed) proving Recent deposits of peat and clay up to 1.5 m thick on weathered Kimmeridge Clay (up to 1 m penetrated).

TL 57 NW/19 to 22 — Great Ouse River Authority Boreholes 38 to 41 [5420 7760 to 5400 7725]
Drilled in 1956 for site investigation. Ground levels 1.7 to 4.1 m above OD.
Shallow boreholes proving up to 9.5 m of Recent deposits (peats and clays with lag gravel at base) on weathered Kimmeridge Clay (up to 1.7 m penetrated).

TL 57 NW/23 — Bridge Road, Ely [5440 7939]
Drilled in 1975 for research. Ground level about OD.

	Thickness m	Depth m
RECENT		
Nordelph Peat	1.7	1.7
Barroway Drove Beds	1.5	3.2
Sand	0.3	3.5
JURASSIC		
Kimmeridge Clay	6.0	9.5

See Gallois, 1976, p.16 for details.

TL 57 NE/2—Bradford Farm, Stuntney [5669 7724]
Drilled for water, probably 19th century. Ground level about
3.5 m above OD.
Ampthill and Kimmeridge clays proved to 39.6 m.
See Whitaker, 1922, p.91 for details.

TL 57 SE/1—BGS Soham Borehole [5928 7448]
Drilled in 1955 for research. Ground level 5.2 m above OD.
See text and Appendix 2 for details.

SHEET TL 58

TL 58 NE/1 to 20—Great Ouse River Authority Boreholes 50 to
63 and 75 to 83 [5741 8513 to 5923 8979]
Drilled in 1950 for site investigation. Ground levels 1.9 to 2.2 m
above OD.
Shallow boreholes (some in river bed) proving Recent deposits of
peat, clay and sand up to 6.4 m thick resting on Kimmeridge
Clay.

TL 58 NE/21 to 32—Great Ouse River Authority Boreholes 14
to 19, 27 to 30, 201 & 202 [5767 8644 to 5931 8999]
Drilled in 1956 for site investigation. Ground levels 0.4 m below
to 4.5 m above OD.
Shallow boreholes proving Recent deposits of peat, clay, sand
and gravelly sand, 4.0 to 9.3 m thick on Kimmeridge Clay.

TL 58 SW/1—Bray's Lane, Ely [5444 8053]
Drilled in 1975 for research. Ground level about 21.3 m above
OD.

	Thickness m	Depth m
CRETACEOUS		
Woburn Sands	3.9	3.9
JURASSIC		
Kimmeridge Clay	5.1	9.0

See Gallois 1976, p.16 for details.

TL 58 SW/2—Market Street, Ely [5428 8043]
Drilled in 1882 for water. Ground level about 22 m above OD.

	Thickness m	Depth m
CRETACEOUS		
Woburn Sands	4.6	4.6
JURASSIC		
Kimmeridge Clay	1.5	6.1

See Whitaker 1922, p.70 for details.

TL 58 SE/1—Prickwillow Bridge [5974 8249]
Drilled in 1960 for site investigation. Ground levels 0.6 m below
to 3.9 m above OD.
Five shallow boreholes proving Recent deposits of peat, silt and
clay up to 7.2 m thick on Kimmeridge Clay.

TL 58 SE/2 to 25—Great Ouse River Authority Boreholes 41 to
49 (1950); 77 and 78 (1950); 31 to 37 (1956) and 43 to 48 (1956)
[5553 8031 to 5376 8470]
Drilled in 1950 and 1956 for site investigation. Ground levels
0.5 m below to 4.3 m above OD.
Shallow boreholes (some in river bed) proving Recent deposits of
peat, clay, sand and gravelly sand 3.5 to 6.3 m thick on Kimmeridge Clay.

TL 58 SE/26—Green Farm, Prickwillow [5989 8230]
Drilled in 1975 for research. About 1.2 m above OD.

	Thickness m	Depth m
RECENT		
Terrington Beds	2.0	2.0
Nordelph Peat	0.7	2.7
Barroway Drove Beds	2.2	4.9
'Lower' Peat	0.2	5.1
Stony sand	0.9	6.0
JURASSIC		
Kimmeridge Clay	8.0	14.0

See Gallois 1976, p.17 for details.

TL 59 SW/1—Copes Hill Farm, Welney [5265 9478]
Drilled in 1975 for research. Ground level about 2.1 m above
OD.

	Thickness m	Depth m
RECENT		
Terrington Beds	c.2.5	c.2.5
Sands: fine- to very fine-grained	9.8	12.3
Gravelly Sand	2.6	14.9
JURASSIC		
? Ampthill Clay	3.1	18.0

See Gallois 1976, p.17 for details.

TL 59 SE/1 to 9—Great Ouse River Authority Boreholes 64 to
66 (1950) and 10 to 13 and 104 (1956) [5948 9021 to 5997 9081]
Drilled in 1950 and 1956 for site investigation. Ground levels
0.5 m below to 4.5 m above OD.
Shallow boreholes (some in river bed) that proved up to 7.8 m of
Recent deposits of peat, clay, sand and gravelly sand.

SHEET TL 67

TL 67 NW/1—Isleham Fen Pumping Station [6280 7881]
Drilled in 1949 for water. Ground level about 0.3 m below OD.

	Thickness m	Depth m
RECENT		
Subsoil and sand	2.4	2.4
CRETACEOUS		
Gault	3.2	5.6
Lower Greensand, undifferentiated	2.9	8.5

Three nearby wells (TL 67 NW/2 to 4) at Prickwillow Road,
Isleham proved the top of the Lower Greensand to be at depths
of 8 to 14 m below OD.

TL 67 NW/5—Cooks Drove Farm, Isleham [6469 7838]
Drilled about 1940 for water. Ground level about 2 m above OD.

	Thickness m	Depth m
PLEISTOCENE		
River Terrace deposits, gravel	4.3	4.3
CRETACEOUS		
Gault	20.7	25.0
Lower Greensand, undifferentiated	3.0	28.0
JURASSIC		
Ampthill and Kimmeridge clays	25.3	53.3

TL 67 NE/4 — Ely-Ouse Water Transfer Scheme: Borehole 12 [6962 7858]
Drilled in 1967 for site investigation. Ground level 3.7 m above OD.

	Thickness m	Depth m
RECENT		
Fine- and medium-grained sand	1.85	1.85
CRETACEOUS		
Lower Chalk	30.83	32.68
Gault	19.54	52.22
Carstone	c.3.1	c.55.3
JURASSIC		
West Walton Beds	c.4.6	59.86

TL 67 NE/6 — Skelton's Farm, Mildenhall [6785 7952]
Drilled in 1901 for water. Ground level about 2.4 m above OD.

	Thickness m	Depth m
PLEISTOCENE		
River Terrace deposits, sand and gravelly sand	5.5	5.5
CRETACEOUS		
Lower Chalk	11.3	16.8

TL 67 NE/7 — Forty Farm, Mildenhall [6522 7922]
Drilled in 1936 for water. Ground level about 1.5 m above OD.

	Thickness m	Depth m
PLEISTOCENE		
River Terrace deposits, sand	4.6	4.6
JURASSIC AND CRETACEOUS		
Gault, Lower Greensand, Kimmeridge and Ampthill clays: no details	91.4	96.0

TL 67 NE/8 — Beck Row Pumping Station [680 772]
Seven boreholes drilled between 1929 and 1964 for water. Ground level 4.5 m above OD.

	Thickness m	Depth m
PLEISTOCENE		
River Terrace deposits, soft yellow sand	1.5	1.5
CRETACEOUS		
Lower Chalk	25.9	27.4
Gault	3.7	31.1

SHEET TL 68

TL 68 NW/1 — Temple Farm, Little Ouse: Borehole A [6292 8685]
Drilled in 1975 for research. Ground level about OD.

	Thickness m	Depth m
RECENT		
Black peaty sand	0.4	0.4
CRETACEOUS		
Woburn Sands	c.2.9	c.3.3
JURASSIC		
Sandringham Sands (Roxham Beds)	c.2.5	c.5.8
Kimmeridge Clay	c.6.7	12.5

See Gallois 1976, p.14 for details.

TL 68 NW/2 — Temple Farm, Little Ouse: Borehole B [6285 8726]
Drilled in 1975 for research. About 0.6 m above OD.

	Thickness m	Depth m
RECENT		
Black peaty sand	0.5	0.5
CRETACEOUS		
Woburn Sands	c.3.9	c.4.4
JURASSIC		
Sandringham Sands (Roxham Beds)	c.2.4	c.6.8
Kimmeridge Clay	c.3.7	10.5

See Gallois 1976, p.15 for details.

TL 68 NE/1 — Ely-Ouse Water Transfer Scheme: Borehole 19 [6930 8534]
Drilled in 1967 for site investigation. Ground level 0.9 m above OD.

	Thickness m	Depth m
RECENT		
Nordelph Peat (1.5 m) on sand and flinty sand	c.8.0	c.8.0
CRETACEOUS		
Lower Chalk	c.13.8	21.79
Gault	c.19.2	c.41.0
Carstone	c.0.8	41.75

TL 68 NE/2 — Ely-Ouse Water Transfer Scheme: Borehole 20 [6928 8618]
Drilled in 1967 for site investigation. Ground level 1.9 m above OD.

	Thickness m	Depth m
RECENT		
Nordelph Peat (1.2 m) on sand and flinty sand	9.22	9.22
CRETACEOUS		
Lower Chalk	14.09	23.31
Gault	18.64	41.95
Carstone	2.49	44.44

TL 68 NE/3 — Ely-Ouse Water Transfer Scheme: Borehole 21
[6923 8683]
Drilled in 1967 for site investigation. Ground level 1.3 m above
OD.

	Thickness m	Depth m
RECENT		
Peaty sand (0.2 m) on sand (4.0 m), clay (4.0 m) and gravelly sand (1.7 m)	9.93	9.93
CRETACEOUS		
Lower Chalk	12.09	22.02
Gault	19.12	41.14
Carstone	0.79	41.93

TL 68 NE/4 — Ely-Ouse Water Transfer Scheme: Borehole 22
[6929 8754]
Drilled in 1967 for site investigation. Ground level 1.1 m above
OD.

	Thickness m	Depth m
RECENT		
Peaty sand (0.5 m) on sand	3.42	3.42
CRETACEOUS		
Lower Chalk	17.91	21.33
Gault	19.91	41.24
Carstone	2.09	43.33

TL 68 NE/5 — Ely-Ouse Water Transfer Scheme: Borehole 23
[6916 8817]
Drilled in 1967 for site investigation. Ground level 2.1 m above
OD.

	Thickness m	Depth m
CRETACEOUS		
Lower Chalk	c.21.8	c.21.8
Gault	c.17.6	c.39.4
Carstone	c.2.6	c.42.0
JURASSIC		
Sandringham Sands	c.2.6	c.44.6
Kimmeridge Clay	c.2.2	46.84

TL 68 NE/11 to 13 — Fenland Flood Protection Scheme
Boreholes 16, 34 and 35 [6918 8948 to 6853 8833]
Drilled in 1959 for site investigation. Ground levels 1.1 to 4.2 m
above O.D.
Shallow boreholes proving Recent deposits of peat and sand up
to 2.3 m thick on Lower Chalk.

TL 68 NE/14 — Shrubhill Farm, Feltwell Anchor [6614 8788]
Drilled in 1975 for research. Ground level about 1.2 m above
OD.

	Thickness m	Depth m
RECENT		
Peaty sand	0.5	0.5
PLEISTOCENE		
Cryoturbated clayey, gravelly sand	1.9	2.4
CRETACEOUS		
Gault	19.9	22.3
Carstone	1.7	24.0

See Gallois 1976, p.14 for details.

TL 68 NE/15 — Corkway Drove, Feltwell Anchor [6763 8972]
Drilled in 1941 for water. Ground level about 0.6 m above OD.

	Thickness m	Depth m
RECENT		
Nordelph Peat	3.1	3.1
CRETACEOUS		
Gault	22.2	25.3
Carstone and ? Sandringham Sands	4.0	29.3
Kimmeridge Clay	0.3	29.6

TL 68 NE/16 — Shrubhill Farm, Feltwell Anchor [6618 8806]
Drilled in 1943 for water. Ground level about 1.8 m above OD.

	Thickness m	Depth m
PLEISTOCENE		
Gravelly sand	3.4	3.4
CRETACEOUS		
Gault	23.4	26.8
Carstone and ? Sandringham Sands	8.0	34.8
JURASSIC		
Kimmeridge Clay	11.2	46.0

TL 68 SW/1 — Shippea Hill Farm, Shippea Hill [6158 8384]
Drilled in 1975 for research. About 6 m above OD.

	Thickness m	Depth m
PLEISTOCENE		
Cryoturbated clayey, gravelly sand	2.1	2.1
JURASSIC		
Kimmeridge Clay	7.9	10.0

See Gallois 1975, p.15 for details.

TL 68 SE/1 — Ely-Ouse Water Transfer Scheme: Borehole 14
[6962 8115]
Drilled in 1967 for site investigation. Ground level 3.8 m above
OD.

	Thickness m	Depth m
Soil and subsoil	0.45	0.45
CRETACEOUS		
Lower Chalk	28.73	29.18
Gault	c.19.0	c.48.2
Lower Greensand, undifferentiated	c.0.7	48.86

TL 68 SE/2 — Ely-Ouse Water Transfer Scheme: Borehole 15
[6949 8213]
Drilled in 1967 for site investigation. Ground level 3.0 m above
OD.

	Thickness m	Depth m
Soil and subsoil	c.1.2	c.1.2
CRETACEOUS		
Lower Chalk	c.27.5	c.28.7
Gault	c.20.4	49.12
Carstone	1.93	51.05

TL 68 SE/3 — Ely-Ouse Water Transfer Scheme: Borehole 16
[6947 8333]
Drilled in 1967 for site investigation. Ground level 1.2 m above
OD.

	Thickness m	Depth m
RECENT		
Nordelph Peat (1.4 m) on fine- and medium-grained sand	6.90	6.90
CRETACEOUS		
Lower Chalk	21.62	28.52
Gault	18.13	46.65
Carstone	0.82	47.47

TL 68 SE/4 — Ely-Ouse Water Transfer Scheme: Borehole 17
[6939 8421]
Drilled in 1967 for site investigation. Ground level 0.9 m above
OD.

	Thickness m	Depth m
RECENT		
Nordelph Peat (1.7 m) on sand with flints	2.59	2.59
CRETACEOUS		
Lower Chalk	c.23.0	c.25.6
Gault	c.19.7	c.45.3
Carstone	c.0.4	45.72

TL 68 SE/5 — Ely-Ouse Water Transfer Scheme: Borehole 18
[6934 8463]
Drilled in 1967 for site investigation. Ground level 1.0 m above
OD.

	Thickness m	Depth m
RECENT		
Nordelph Peat (1.2 m) on sand	3.81	3.81
CRETACEOUS		
Lower Chalk	20.80	24.61
Gault	c.19.3	c.43.9
Carstone	c.2.0	45.92

TL 68 SE/8 — Poplar Farm, Mildenhall [6657 8092]
Drilled in 1952 for water. Ground level 1.5 m above OD.

	Thickness m	Depth m
PLEISTOCENE		
Sand	4.3	4.3
CRETACEOUS		
Lower Chalk	6.1	10.4
Gault	18.3	28.7
Lower Greensand (probably Carstone)		just seen

SHEET TL 69

TL 69 NW/1 — A10 Improvement Scheme: Borehole 3 [6132
9569]
Drilled in 1970 for site investigation. Ground level 2.6 m above
OD.

	Thickness m	Depth m
RECENT		
Black peaty clay	3.0	3.0
JURASSIC		
Kimmeridge Clay	15.0	18.0

TL 69 NW/8 — A10 Improvement Scheme: Borehole 16 [6152
9505]
Drilled in 1970 for site investigation. Ground level 7.4 m above
O.D.
Chalky-Jurassic till seen to 2.90 m.

TL 69 NW/9 and 10 — Great Ouse River Authority Boreholes 22
and 23 [6011 9510 and 6013 9512]
Drilled in 1956 for site investigation. Ground levels 4.2 and
0.3 m above OD.
Shallow boreholes that proved Recent deposits of Terrington
Beds silts (up to 1.8 m) on Nordelph Peat (up to 2.0 m) on thin
sands on weathered Kimmeridge Clay.

TL 69 NE/9 — Severals House, Methwold [6921 9639]
Drilled in 1920 for Kimmeridgian oil shale. Ground level about
3 m above OD.

	Thickness m	Depth m
RECENT		
Nordelph Peat	0.6	0.6
CRETACEOUS		
Lower Chalk	c.11.6	c.12.2
Gault and ? Carstone	c.18.2	c.30.4
JURASSIC		
Sandringham Sands (Roxham Beds)	c.0.9	c.31.3
Kimmeridge Clay	c.45.8	c.77.1
Ampthill Clay	c.20.4	c.97.5
West Walton Beds	c.7.7	c.105.2
Oxford Clay	c.39.6	144.78
Kellaways Beds	1.37	146.15
Great Oolite 'Series' and ? Cornbrash	3.04	149.19
Middle Lias	11.74	160.93
Lower Lias	43.28	204.21

See Pringle 1923, p.129-132 and present text for details.

TL 69 SW/1 to 8 — Great Ouse River Authority Boreholes 67 to
74 [6019 9117 to 6123 9350]
Drilled in 1950 for site investigation. River bed levels 1.4 to
3.0 m below OD.
Shallow boreholes drilled in the river bed that proved Recent
deposits of peat, clay and gravelly sand 1.1 to 2.9 m thick on
Kimmeridge Clay.

TL 69 SW/9 — A10 Improvement Scheme: Brandon Creek
Bridge [6072 9172]
Drilled in 1966 for site investigation. Ground levels 0.6 m below
to 4.1 m above OD.
Eight shallow boreholes that proved Recent deposits of peat, clay
and sand and gravel up to 14.6 m thick on Kimmeridge Clay.

TL 69 SW/10 to 21 — A10 Improvement Scheme Boreholes 1, 2,
14 and 17 to 25 [6087 9188 to 6164 9440]
Drilled in 1970 for site investigation. Ground levels 2.0 m below
to 3.2 m above OD.

Shallow boreholes that proved Recent deposits of peat, clay and sand and gravel ranging from 1.8 m thick (near Southery) to 7.9 m thick (near Brandon Creek) resting on Kimmeridge Clay.

TL 69 SW/22 to 37 — Great Ouse River Authority Boreholes 1 to 9, 24 to 26 and 101 to 105 [6061 9195 to 6032 9482]
Drilled in 1956 for site investigation. Ground levels 1.2 m below to 4.4 m above OD.
Shallow boreholes that proved Recent deposits of peat, clay and gravelly sand 1.8 to 10.8 m thick on Kimmeridge Clay.

TL 69 SW/38 — Upgate Street, Southery [6208 9483]
Drilled in 1975 for research. Ground level about 1.8 m above OD.

	Thickness m	Depth m
RECENT		
Black stony sand	1.35	1.35
JURASSIC		
Sandringham Sands (Roxham Beds)	3.6	4.95
Kimmeridge Clay	3.55	8.50

See Gallois 1976, p.13 for details.

TL 69 SW/39 — Decoy Farm, Southery [6490 9463]
Drilled in 1975 for research. Ground level about OD.

	Thickness m	Depth m
RECENT		
Peaty sand	0.5	0.5
CRETACEOUS		
Carstone	2.7	3.2
JURASSIC		
Sandringham Sands (Roxham Beds)	c.2.9	c.6.1
Kimmeridge Clay	c.4.9	11.0

See Gallois 1976, p.13 for details.

TL 69 SW/40 — Decoy Farm, Southery [6485 9482]
Drilled in 1920 for Kimmeridgian oil shale. Ground level about 1.5 m above OD.
Cretaceous sequence probably as TL 69 SW/39: Kimmeridge Clay proved from 12.2m to 42.8 m.
See Pringle 123, p.127 for details.

TL 69 SE/2 — Methwold Common, Methwold [678 941]
Drilled in 1920 for Kimmeridgian oil shale. Ground level about 2.1 m above OD.

	Thickness m	Depth m
RECENT		
Peat, sand and flinty sand	9.1	9.1
CRETACEOUS		
Gault and ? Carstone	11.0	20.1
JURASSIC		
Sandringham Sands (Roxham Beds)	0.4	20.5
Kimmeridge Clay	42.6	63.1
Ampthill Clay	21.6	84.7
West Walton Beds	5.7	90.4
Oxford Clay	31.1	121.5

See Pringle 1923, pp.128 – 129 for details and Gallois and Cox 1977, pp.220 – 221 for revised classification of Jurassic.

SHEET TL 78

TL 78 SW/1 — Lakenheath Borehole [748 830]
Drilled by Superior Oil (UK) Ltd in 1964 for hydrocarbon exploration. Ground level about 7.6 m above OD.
See text for details.

TL 78 SE/1 — Ely-Ouse Water Transfer Scheme: Borehole 13 [7954 8001]
Drilled in 1967 for site investigation. Ground level 3.2 m above OD.

	Thickness m	Depth m
RECENT		
Sand with flints	2.97	2.97
CRETACEOUS		
Lower Chalk	26.59	29.56
Gault	17.53	47.09
Carstone	2.79	49.88

APPENDIX 2

Summary log of the BGS Soham Borehole

The Soham Borehole was drilled in 1955 to investigate the Palaeozoic and Mesozoic sequence in an area where geophysical evidence (Bullard and others, 1940) had suggested the basement to occur at relatively shallow depth. The borehole was sited [5928 7448] near Qua Fen Common on First Terrace gravels at a height of 5.2 m above OD. Below a depth of 4.6 m the borehole was continuously cored and a suite of geophysical logs was run by Schlumberger Limited. The borehole was supervised and logged by Mr S. C. A. Holmes who has published a summary of the sequence (*in* Worssam and Taylor, 1969, pp.7-8). Additional collecting and notes were made by Messrs H. G. Dines and R. V. Melville.

The borehole summary given below is based on a re-examination of the specimens by the present author and Dr Ivimey-Cook (Liassic, Bathonian and Bajocian faunas), Dr Cox (Upper Jurassic faunas) and Mr Morter (Albian faunas). Dr Ivimey-Cook and Dr Cox have made the zonal interpretations of the Lower and Upper Jurassic respectively, and Mr Morter and the author have classified the Gault.

BOREHOLE SUMMARY

	Thickness m	Depth m
PLEISTOCENE		
First Terrace, sand and gravel (rock cuttings only)	3.0	3.0
CRETACEOUS		
Upper Gault		
Soft mudstones aod calcareous mudstones (rock cuttings only)	1.6	4.6
Beds G 11 to G 14: mudstones and calcareous mudstones; shelly in part with common ammonites including *Euhoplites alphalautus, E. inornatus, Hysteroceras, Mortoniceras* and *Semenovites*, bivalves including *Birostrina sulcata* and '*Inoceramus' anglicus*, and *Nielsenicrinus*	c.12.1	c.16.7
Lower Gault		
Beds G 1 to G 10: mudstones and calcareous mudstones; shelly in part with rare ammonites including *Euhoplites* cf. *opalinus*, common bivalves including *Bakevellia, Birostrina concentrica* and *Nucula*; sandy and pebbly at base with phosphatised burrowfills penetrating the top of the underlying bed	c.5.8	22.45
Carstone		
Sandstone; soft, pebbly, oolitic (limonite); bioturbated throughout, barren; base in core loss	c.0.7	c.23.2
Woburn Sands		
Sandstone, pebbly, glauconitic and limonitic; calcite cemented in part; pebbles include oolitic ironstone (? from Sandringham Sands), vein quartz and lydite. Fragmentary sample recovered as cavings	c.1.3	c.24.5

	Thickness m	Depth m
JURASSIC		
West Walton Beds		
Cardioceras tenuiserratum Zone		
Siltstone, clayey; pale grey, calcareous and in part calcareously cemented, passing down into heavily bioturbated clayey silts rich in foraminifera and containing scattered oncoids (algal pellets) that become more common with depth; passing down into	0.89	25.40
Interbedded oncoid-rich (>50%) pale grey calcareous mudstone and oncoid-rich shell-fragmental limestone	1.57	26.97
Mudstone, silty and clayey siltstone; calcareously cemented in part; sparsely shelly with *Cardioceras tenuiserratum* (Oppel); bioturbated junction with	1.10	28.07
Cardioceras densiplicatum Zone		
Mudstone, silty; pale grey becoming darker with depth, perisphinctid ammonites common at 30.5 to 35.2 m; silty bioturbated oyster-rich bed at base with large serpulid-encrusted and intensely bored oysters resting on an irregular, burrowed surface of Oxford Clay	7.29	35.36
Upper Oxford Clay		
Quenstedtoceras mariae Zone		
Mudstone, soft; uniformly pale grey, slightly silty, calcareous; locally more calcareous and with sub-conchoidal fracture; passing down into medium and pale grey mudstones and silty mudstones; cemented to form thin tabular limestones at two depths; sparsely shelly but with pyritised *Pinna* locally abundant; other bivalves include arcids, *Gryphaea, Modiolus, Nicaniella (Trautscholdia)*, nuculoids including *Dacryomya* and *Mesosaccella, Oxytoma, Protocardia* and *Thracia*: rare small belemnites (*Hibolites*), *Dicroloma* and abundant pyritised trails and burrows also present; crustacean debris occurs at several levels; small pyritised ammonites, many uncrushed, including *Cardioceras, Euaspidoceras, Peltoceras* and *Quenstedtoceras* occur throughout; the presence of *Cardioceras praecordatum* Douvillé at 35.56 and 37.80 m indicates the *praecordatum* Subzone; *C. scarburgense* (Young & Bird) at 42.69 and 47.30 m to 47.55 m and *Quenstedtoceras omphaloides* (J. Sowerby) at 48.79 and 48.92 m indicate the *scarburgense* Subzone; the junction of the two subzones has been taken at about 41.30 m at the top of a thin cementstone which marks a general downward change to slightly more silty mudstones; the zonal index *Quenstedtoceras mariae* (d'Orbigny) was recorded in both subzones	14.50	49.86
Middle Oxford Clay		
Quenstedtoceras lamberti Zone		
Mudstone, very silty; bioturbated, medium grey and pale, faintly greenish grey, calcareous; becoming cemented to form a soft muddy limestone (Lamberti Limestone) at 49.91 to 50.57 m; very shelly at 49.86 to 49.91 m with common thick shelled *Gryphaea* (many of which are water-worn, bored and		

	Thickness m	Depth m

encrusted by foraminifera and serpulids) and with pyritised, crushed cardioceratids; moderately shelly below 49.91 m with scattered bivalves and small partially pyritised ammonites, common shell debris and abundant pyritised trails and burrows; bivalve fauna similar to that of the Upper Oxford Clay but with *Pinna* now rare and *Bositra buchii* (Roemer) also present; the serpulid *Genicularia vertebralis* (J. de C. Sowerby) is present at 51.44 m together with nests of *Nicaniella* (*Trautscholdia*); ammonites include *Alligaticeras sp.* at 50.47 and 51.21 m, *Kosmoceras* (*Spinikosmokeras*) *sp.* at 51.28 m and *Quenstedtoceras sp.* at 51.31 ; base of zone taken at a *Gryphaea*-rich band that approximately marks a downward change to less silty mudstone ... 1.80 ... 51.66

Peltoceras athleta Zone (pars)
Mudstone, silty and slightly silty; pale grey and pale, faintly brownish grey, calcareous; passing into soft muddy limestone at several levels; moderately fossiliferous with partially pyritised bivalves abundant at many levels; common *Bositra* and *Nicaniella*, encrusted *Gryphaea*, nuculoids, *Oxytoma*, *Thracia*, *Dicroloma*, common *Genicularia*, partially pyritised wood; rare crushed ammonites include *Hecticoceras sp.* at 57.58 m and *Kosmoceras sp.* at 58.14m and 58.37 m; base of Middle Oxford Clay taken at the upper limit of slightly bituminous mudstones ... 7.32 ... 58.98

Lower Oxford Clay
Peltoceras athleta Zone (pars)
Mudstone, silty and slightly silty; pale grey and pale, faintly brownish grey, calcareous; passing into soft muddy limestone at several levels; interbedded with fissile, slightly brownish grey slightly bituminous mudstones crowded with tiny flakes of bivalve shell and spat; moderately shelly (calcareous beds) to very shelly (bituminous beds) with fauna composed mostly of crushed, partially pyritised, ammonites and bivalves; *Kosmoceras* including *K.* (*Spinikosmokeras*) *sp.* at 59.03 m. *K.* (*Hoplikosmokeras*) *fibuliferum* Buckman at 62.33 m, *K.* (*H*). *phaeinum* Buckman at 66.35 m, *K.* (*Spinikosmokeras*) *acutistriatum* Buckman at 67.54 and 69.37 m, and *K.* (*S.*) cf. *rimosum* (Quenstedt) at 69.74 m; bivalve fauna similar to bed above but with *Bositra buchii*, *Meleagrinella* and *Mesosaccella* especially common; *Grammatodon*, rare *Gryphaea*, *Isocyprina*, *Thracia*, cerithiid gastropods, locally common *Genicularia vertebralis* and rare belemnites also present ... 11.35 ... 70.33

Erymnoceras coronatum Zone
Mudstone, weakly calcite-cemented; pale grey, extremely shelly with abundant partially pyritised nuculoid bivalves; shell plaster with common wood fragments and crushed *Binatisphinctes comptoni* Pratt at 70.33 to

	Thickness m	Depth m

70.35 m (the Comptoni Bed of the East Midlands); other ammonites include *Erymnoceras sp.* at 71.02 m, *Kosmoceras* (*Spinikosmokeras*) cf. *castor* (Reinecke) at 70.41 and 70.94 m and *K.*(*Zugokosmokeras*) *grossouvrei* Douvillé at 71.12 and 71.20 m; resting with bioturbated junction on ... 0.87 ... 71.20
Rhythmic, thinly bedded alternations of blocky, pale grey, moderately shelly, calcareous mudstone and fissile, brownish grey, very shelly, bituminous mudstone mostly with pyritised fossils; traces of lamination preserved in the bituminous beds but mostly destroyed by bioturbation adjacent to contacts with the calcareous mudstones; non-ammonite fauna as zone above with nuculoids especially common in the bituminous beds; ammonites include *K.* (*Hoplikosmokeras*) *pollucinum* Teisseyre at 71.32, 72.89 and 73.20 m, *Kosmoceras* (*Zugokosmokeras*) *grossouvrei* at 71.93 to 72.21 m, *K.* (*Z.*) *obductum* (Buckman) at 73.66, 74.22 and 74.58 m and *Erymnoceras sp.* at 74.20 m; zonal boundary taken on the basis of the ammonites ... c.3.9 ... c.75.1

Kosmoceras jason Zone
Rhythmic alternations of calcareous mudstone and very shelly bituminous mudstone; *Kosmoceras* (*Gulielmites*) *jason* (Reinecke) at 75.19 to 76.28 m; bivalve and other fauna as in the *coronatum* Zone, except at base where extremely shelly mudstone, crowded with crushed ammonites and with gryphaeid oysters and belemnites, including *Cylindroteuthis* and *Hibolites*, rests with an irregular, apparently erosive base on the bed below ... c.1.7 ... 76.78

Sigaloceras calloviense Zone (pars)
Mudstone, bituminous, finely laminated, bioturbated, brownish grey; very shelly with abundant crushed *Kosmoceras* and rare *Sigaloceras* (*Catasigaloceras*) *enodatum* (Nikitin); bivalves including *Gryphaea*, *Nicaniella* (*Trautscholdia*) and *Pinna*; *Procerithium*, belemnites and fish debris also present; passing rapidly down into shelly, very bioturbated, silty mudstone and then with bioturbation into ... 0.11 ... 76.89

Kellaways Rock
Sigaloceras calloviense Zone (pars)
Siltstone, greyish brown, weakly calcite-cemented; intensely bioturbated with burrowfills of pale grey and brownish grey clay throughout; very shelly with very common *Gryphaea* (*Bilobissa*) *dilobotes* Duff and common belemnites; ammonite fragments common including *Proplanulites* and *Sigaloceras*; larger fossils encrusted with coarsely crystalline pyrite; passing down with bioturbation into ... 0.30 ... 77.19

Kellaways Clay
Macrocephalites macrocephalus Zone (pars)
Mudstone, greyish brown, silty and with ramifying network of pyritic cement (?burrowfills) occurring throughout; barren

	Thickness m	Depth m

except for rare *Meleagrinella*; resting with bioturbated junction but with little lithological change, and a single, dense, phosphatic concretion on | 0.17 | 77.36

Upper Cornbrash
Macrocephalites macrocephalus Zone (pars)
Mudstone; greyish brown with rare burrowfill wisps of rotted limonitised shell debris, these increase rapidly with depth and are an important constituent below 77.52 m; sparsely shelly in upper part but with *Lopha marshii* (J. Sowerby) at 77.52 m and *Macrocephalites sp.* at 77.55 m; becoming more silty and shelly with depth with small oysters, paired valves of *Goniomya*, small pectinids, a large *Serpula*, pyritised trails and burrow concentrations of small bivalves; irregular base with silty mudstone infilling hollows and borings (cf. *Trypanites*) in underlying bed | 0.31 | 77.67

Limestone, densely cemented, silty with rare concentrations of limonitised shell debris, non-calcareous matrix lithologically similar to bed above; sparsely shelly with *Entolium* and other bivalves; becoming very silty and recrystallised in lowest part with limonite-coated cavities; large solitary coral and angular phosphatic pebble close to base; intensely bioturbated at base with shelly silt infilling burrows and ?dessication cracks in underlying bed | 0.82 | 78.49

Upper Estuarine 'Series'
Mudstone, finely laminated pale greenish grey; preserved only as lithorelics between parallel sided (?dessication) cracks and burrows infilled by overlying bed; planar base | 0.02 | 78.51

Limestone, pale, slightly greenish grey becoming slightly pinkish grey at base; dense, silty; barren except for possible oyster fragments; passing down into | 0.10 | 78.61

Siltstone, mottled medium grey, pale, slightly greenish grey and slightly pinkish grey; intensely bioturbated but with traces of irregular lamination preserved; almost barren but becoming shelly with depth with *Liostrea, Modiolus, Placunopsis* and *Vaugonia*; irregular contact with | 0.23 | 78.84

Oyster lumachelle with dark grey clay matrix; composed almost entirely of *Liostrea hebridica* (Forbes) in upper part; weakly cemented at 79.66 m but mostly collapsing to shelly brash; passing down into very shelly, silty mudstone and siltstone below 79.71 m with much shell dust and debris and a more diverse fauna including *Anisocardia, Corbulomima, Modiolus, Praemytilus, Pseudolimea* and *Procerithium*; very irregular bioturbated junction with | 1.19 | 80.03

Lower Estuarine 'Series'
Siltstone, coarse, and very fine-grained sandstone; mostly pale and very pale grey but darker at several levels due to comminuted plant debris; brownish grey podzol down to 80.2 m; carbonised vertical plant rootlets ubiquitous in upper part; coalified wood fragments abundant at several levels; traces of

	Thickness m	Depth m

lamination (broken by rootlets) at one level; rare thin dark grey carbonaceous silty mudstone bands; becoming tougher, more uniformly silty and clayey below 86.7 m with scattered nests of sphaerosiderite pellets giving rise to purplish red and yellowish brown (limonitic) staining at 90.22 to 90.30 m; becoming finely sandy and bioturbated at base | c.10.8 | c.90.8

LOWER LIAS
Tragophylloceras ibex Zone
Mudstone, deeply weathered, pale grey (leached) with traces of purplish red (?incipient sphaerosiderite formation in former soil profile at 90.9 m), and pale yellow staining mechanically weathered with lithorelics of mudstone set in a clay matrix and with common striated slip surfaces; rapid downward change into | c.0.9 | 91.72

Mudstone, unweathered pale and medium grey slightly silty and silty, almost barren; rare ammonites include *Beaniceras sp.* at 94.56 m; pyrite trails and knots common; tabular limonite-coated clay ironstone at 94.97 m; thin interbeds of pyritic shelly mudstone; passing down into | c.3.6 | c.95.3

Siltstone and silty mudstone, medium grey; shelly with common large coarsely-ribbed pectinids, and belemnites; tabular clay ironstone at 95.71 m; passing down into | c.2.2 | 97.48

Limestone, dense shelly brash set in mudstone matrix, interbedded with pale grey slightly silty mudstones; oysters, pectinids, belemnites and rhynchonellids common in the limestones; densely cemented limestone at base with limonitised shell debris; sharp base | 1.60 | 99.08

Siltstone, medium and pale grey with a few thin interbeds of silty mudstone; calcareously cemented at several levels; moderately to sparsely shelly with common myid bivalves in life position; patches of pale brown limonitic cement at several levels; densely ferruginously cemented shelly (oysters) limestone lens at 103.33 m; clay ironstone burrowfill at 106.38 m; passing down into | 7.30 | 106.38

Mudstone, silty and highly silty; brown purplish red; weak ferruginous cement at several levels; sparsely shelly but with a few thin very shelly bands with common bivalves and rare ammonites; *Tragophylloceras sp.* at 107.95 m, *Liparoceras sp.* at 119.48 m, *Chondrites* and other bioturbation at several levels; some lamination; passing down into | 18.74 | 125.12

Mudstone, silty, and siltstone; medium grey becoming brownish grey with depth; shelly throughout with much pyritised shell brash; ammonites, belemnites, oysters and rhynchonellids especially common; bioturbated throughout; *Acanthopleuroceras valdani* (d'Orbigny) abundant between 126.19 and 126.87 m; passing down with bioturbation into | c.3.3 | c.128.4

Uptonia jamesoni Zone
Mudstone, pale grey, silty, interbedded with brashy, weakly cemented shelly mudstone and

	Thickness m	Depth m

almost barren weakly cemented siltstone; bioturbated throughout; shelly bands particularly rich in bivalves, brachiopods and ammonites including *Uptonia sp.* at 128.42 m, common *U. bronni* (Roemer) from 129.54 to 130.30 m and a large oxycone at 130.89 m; passing down into — c.3.4 — 131.83

Siltstone, brownish grey, oolitic (limonite) with ooid concentration increasing with depth until densely oolitic mudstone in lower part; bioturbated throughout with burrowfills of pale grey silt and muddy silt; small limonite-coated pebbles common in lower part; sparsely shelly but with bivalves, brachiopods and ammonites locally common; *Apoderoceras sp.* at 132.71 m; passing down into — 1.09 — 132.92

Pebble bed; grey silty mudstone packed with angular fragments of red and green mudstone and well rounded pebbles of sandstone, dolomite and vein quartz (all of Triassic derivation); very irregular base at 133.02 to 133.07 m with well preserved burrows infilled with oolitic (limonitic) siltstone extending down to at least 133.77 m — 0.15 — 133.07

TRIASSIC

Siltstone, pale slightly greenish grey inter-bedded with finely micaceous fine-grained sandstone; patchy dolomitic cement at 134.4, 134.7 and 136.4 m; becoming mottled pale greyish green and dull reddish brown below 135.6 m; pellet conglomerate with small, angular mudstone pebbles at 135.9 to 136.3 m; predominantly sandy from 136.3 to 139.6 m with dolomitic cement at 136.3 to 136.6 m and nodular dolomite concretions (? replacing anhydrite) at several levels between 137.8 and 139.6 m; passing down into — 6.53 — 139.60

Mudstone, silty and muddy siltstone; dull red with some pale green banding and mottling; several thin sandstone interbeds; traces of horizontal lamination; sharp base — 6.70 — 146.30

Sandstone, coarse-grained, pebbly with minor thin beds of red silty mudstone; pebbles most-ly Triassic sandstone and mudstone and well-rounded vein quartz; pebbles becoming com-moner and larger with depth below 153.6 m until very coarse conglomerate at base includes angular small boulders of Triassic mudstone; very irregular, presumed erosive base — 10.70 — 157.00

Sandstone, mottled dull red and greyish green fine-grained becoming pebbly below 162.1 m with large number of rotted limonite pebbles (hence yellow and brown colouring); patchy dolomitic cement at several levels; pebble bed at base and sharp junction with — 5.66 — 162.66

Mudstone, silty, mottled red and green, crowded with angular (joint and bedding bounded) fragments of dark purplish red mudstone from underlying bed; highly ir-regular junction with Palaeozoic mudstones — 0.74 — 163.4

	Thickness m	Depth m

SILURIAN – DEVONIAN

Mudstone, silty in part; dark purplish red with more orange-red (Triassic staining) patches down to 173.7 m and with pale greyish green reduction patches in body of rock and adjacent to joints; intensely fractured *in situ* with open joints and bedding planes (ruckled in upper part) largely infilled with dolomite; passing down into — 3.30 — 166.70

Mudstone, pale greenish grey with dull purplish and brownish red staining adjacent to joints and bedding planes and, at a few levels, permeating the whole rock; staining becomes less with depth and below about 184.4 m staining is restricted to dark reddish brown coating on major joints; matrix becomes progressively darker at approximately the same depth, passing down into — 20.80 — 187.50

Mudstone, medium and dark grey; rhythmic throughout with graded bedding in units mostly 1 to 10 cm thick in which sparsely shelly, medium grey, silty mudstone passes up into barren, dark grey mudstone; base of units commonly erosional on underlying mudstone with sole structures and common burrowing, including *Chondrites;* thin beds (mostly 4 to 15 cm thick) of micaceous, finely planar, cross- and ripple-laminated siltstones become common below 187.5 m; ripple-drift bedding and small washouts present in some bands but mostly planar lamination; contorted bedding (loading or slumping) in siltstone at 195.1 m; possible water-escape structures at 193.77 m; barren to sparsely shelly throughout but with widely spaced bedding planes crowded with poorly preserved bivalves and ostracods; 15 cm-thick bed of very shelly crystalline limestone composed largely of gastropods at 203.9 m; thick siltstone at 210.6 to 210.9 m with ripple-drift lamination and thin films of dull purplish red staining adjacent to joints — 23.4 — 210.9

Mudstone, grey and slightly greenish grey with all joints stained reddish brown; lithologies as bed above; mudstones and some siltstones becoming irregularly mottled purplish red-brown and pale green below 221.0 m; contorted bedding (?loaded ripples) at 221.0 and 226.8 m — 31.26 — 242.16

Final depth — — 242.16

REFERENCES

ANDERTON, R., BRIDGES, P. H., LEEDER, M. R: and SELLWOOD, B. W. 1979. *A dynamic stratigraphy of the British Isles.* (London: Allen & Unwin.)

ANON 1955. Gravity survey overlay map. Sheet 16. *Geol. Surv. G.B.*

— 1959. Aeromagnetic map of Great Britain. Sheet 2: England and Wales. *Geol. Surv. G.B.*

ARKELL, W. J. 1933. *The Jurassic System in Great Britain.* (Oxford: Clarendon Press.)

— 1937. Report on ammonites collected at Long Stanton, Cambs., and the age of the Ampthill Clay. *Summ. Prog. Geol. Surv. G.B.* for 1935, Pt. II, 64–88.

— 1947. The geology of the country around Weymouth, Swanage, Corfe, and Lulworth. *Mem. Geol. Surv. G.B.*.

ASTBURY, A. K. 1958. *The Black Fens.* (Cambridge: Golden Head Press.)

BADEN-POWELL, D. F. W. 1934. On the marine gravels at March, Cambridgeshire. *Geol. Mag.*, Vol. 71, 193–219.

— 1948. The chalky boulder clays of Norfolk and Suffolk. *Geol. Mag.*, Vol. 85, 279–96.

BARRET, W. H. 1963. *Tales from the Fens.* PORTER, E. (editor). (London: Routledge and Kegan Paul.)

— 1964. *More tales from the Fens.* PORTER, E. (editor). (London: Routledge and Kegan Paul.)

BELL, F. G. 1970. Late Pleistocene floras from Earith, Huntingdonshire. *Phil. Trans. R. Soc.*, Ser. B, Vol. 258, 347–78.

BLACK, M. 1953. The constitution of the Chalk. *Proc. Geol. Soc. London,* No. 1499, lxxxi-vi.

— 1972–1975. British Lower Cretaceous coccoliths. 1. Gault Clay. *Palaeontogr. Soc. Monogr.* 1–142.

BLAKE, J. F. 1875. On the Kimmeridge Clay of England. *Q. J. Geol. Soc. London,* Vol. 31, 196–233.

— and HUDLESTON, W. H. 1877. On the Corallian rocks of England. *Q. J. Geol. Soc. London,* Vol. 33, 260–405.

BONNEY, T. G. 1872a. On the section exposed at Roslyn Hill pit, Ely. *Proc. Cambridge Phil. Soc.*, Vol. 2, 268.

— 1872b. Notes on the Roslyn Hill Clay pit. *Geol. Mag.*, Vol. 4, 403–405.

— 1875. *Cambridgeshire geology.* (London: Reighton Bell & Co.)

BOSWELL, P. G. H. 1927. On the distribution of purple zircon in British sedimentary rocks. *Mineral. Mag.*, Vol. 21, 310–317.

BREISTROFFER, M. 1947a. Sur L'age des gres verts de Cambridge (Angleterre). *Soc. Géol. Fr.*, Sec. Dec. 1946, 309–312.

— 1947b. Sur les zones d'ammonites dans l'Albien de France et l'Angleterre. *Trav. Lab. Geol. Univ. Grenoble*, Vol. 26, 17–104.

— 1965. Vues sur les zones d'ammonite de l'Albien. *Mém. Bur. Rech. Géol. Minières.*, No. 34, 311–312.

BRIGHTON, A. G. 1938. The Mesozoic rocks of Cambridgeshire. 6–15 in *A scientific survey of the Cambridge district.* DARBY, H. C. (editor). (London: British Association for the Advancement of Science.)

BRISTOW, C. R. In press. The geology of the country around Bury St Edmunds. *Mem. Geol. Surv. G.B.*.

— and Cox, F. C. 1973. The Gipping Till: a reappraisal of East Anglian glacial stratigraphy. *J. Geol. Soc. London.*, Vol. 129, 1–37.

BULLARD, E. C., GASKELL, T. F., HARLAND, W. B. and KERR-GRANT, C. 1940. Seismic investigations on the Palaeozoic floor of east England. *Phil. Trans. R. Soc.*, Ser. A, Vol. 239, 29–94.

BURNABY, T. P. 1962. The palaeoecology of the foraminifera of the Chalk Marl. *Palaeontology,* Vol. 4, 599–608.

BUTLER, D. E. 1981. Marine faunas from concealed Devonian rocks of southern England and their reflection of the Frasnian transgression. *Geol. Mag.*, Vol. 118, 679–697.

CALLOMON, J. H. 1968. The Kellaways Beds and the Oxford Clay. 264–290 in *The geology of the East Midlands.* SYLVESTER-BRADLEY, P. C. and FORD, T. D. (editors) (Leicester: Leicester University Press.)

CAMERON, A. G. C. 1892. The Fuller's Earth Mining Co. at Woburn Sands. *Geol. Mag.*, Vol. 9, 469.

CARTER, D. J. and HART, M. B. 1977. Aspects of mid-Cretaceous stratigraphical micropalaeontology. *Bull. Br. Mus. (Nat. Hist.) Geol.*, Vol. 29, 1–135.

CASEY, R. 1960. P.48 in *Summ. Prog. Geol. Surv. G.B.* for 1959.

— 1961a. Geological age of the Sandringham Sands. *Nature, London,* No. 4781, Vol. 190, 1100.

— 1961b. The stratigraphical palaeontology of the Lower Greensand. *Palaeontology,* Vol. 3, 487–621.

— 1962. The ammonites of the Spilsby Sandstone, and the Jurassic–Cretaceous boundary. *Proc. Geol. Soc. London,* No. 1598, 95–100.

— 1967. The position of the Middle Volgian in the English Jurassic. *Proc. Geol. Soc. London,* No. 1640, 128–133.

— 1973. The ammonite succession at the Jurassic–Cretaceous boundary in eastern England. 193–266 in The Boreal Lower Cretaceous. CASEY, R. and RAWSON, P. F. (editors) *Geol. J. Spec. Issue* No. 5.

— and GALLOIS, R. W. 1973. The Sandringham Sands of Norfolk. *Proc. Yorkshire Geol. Soc.*, Vol. 40, 1–22.

CHANDLER, R. J. 1972. Lias Clay: weathering processes and their effects on shear strength. *Géotechnique*, Vol. 22, 403–431.

CHATWIN, C. P. 1948. *British regional geology: East Anglia and adjoining areas* (3rd Edition). (London: HMSO for Geological Survey and Museum.)

CHROSTON, P. N. 1985. A seismic refraction line across Norfolk. *Geol. Mag.*, Vol. 122, 397–401.

— and SOLA, M. 1975. The sub-Mesozoic floor in Norfolk. *Bull. Geol. Soc. Norfolk.*, Vol. 27, 3–19.

CLARK, J. G. D. 1933. Report on an Early Bronze Age site in the south-eastern fens. *Antiquaries J.*, Vol. 13, 266–296.

— and GODWIN, H. 1962. The Neolithic in the Cambridgeshire fens. *Antiquity*, Vol. 36, 10–23.

— — GODWIN, M. E. and CLIFFORD, M. H. 1935. Report of recent excavations at Peacock's Farm, Shippea Hill, Cambridgeshire. *Antiquaries J.*, Vol. 15, 284–319.

COOKSON, I. C. and HUGHES, N. F. 1964. Microplankton from the Cambridge Greensand (Mid-Cretaceous). *Palaeontology,* Vol. 7, 37–59.

COPE, J. C. W. 1967. The palaeontology and stratigraphy of the lower part of the Upper Kimmeridge Clay of Dorset. *Bull. Br. Mus. (Nat. Hist.) Geol.*, Vol. 15, 3–79.

— 1978. The ammonite faunas and stratigraphy of the upper part of the Upper Kimmeridge Clay of Dorset. *Palaeontology*, Vol. 21, 469–533.

— DUFF, K. L., PARSONS, C. F., TORRENS, H. S., WIMBLEDON, W. A. and WRIGHT, J. K. 1980. A correlation of Jurassic rocks in the British Isles. Part Two: Middle and Upper Jurassic. *Spec. Rep. Geol. Soc. London*, No. 15.

COX, B. M. and GALLOIS, R. W. 1981. The stratigraphy of the Kimmeridge Clay of the Dorset type area and its correlation with some other Kimmeridgian sequences. *Rep. Inst. Geol. Sci.*, No. 80/4.

COX, F. C., POOLE, E. G. and WOOD, C. J. *in press*. The geology of the country around Norwich. *Mem. Geol. Surv. G.B.*.

DANSGAARD, W., JOHNSEN, S. J., CLAUSEN, H. B. and LANGWAY, C. C. 1971. Climatic record revealed by the Camp Century ice core. 37–56 in *Late Cenozoic ice ages*. (New Haven, Conn.: Yale University Press.)

DARBY, H. C. 1940a. *The draining of the fens*. (Cambridge: Cambridge University Press.)

— 1940b. *The medieval fenland*. (Cambridge: Cambridge University Press.)

— 1971. *The Domesday geography of eastern England*. (Cambridge: Cambridge University Press.)

DEAN, W. T., DONOVAN, D. T. and HOWARTH, M. K. 1961. The Liassic ammonite zones and subzones of the north-west European province. *Bull. Br. Mus. (Nat. Hist.) Geol.*, Vol. 4, No. 10, 435–505.

DE RANCE, C. E. 1868. On the Albian or Gault of Folkestone. *Geol. Mag.*, Vol. 5, 163–171.

DIXON, E. E. L. 1933. The Gault of Cambridgeshire. *Summ. Prog. Geol. Surv. U.K.* (for 1932), 78–81.

DODSON, W. 1665. *The design for the perfect draining of the great level of the fens (called Bedford Level) etc.* (London: R. Wood.)

DONN, W. L., FARRAND, W. R. and EWING, M. 1962. Pleistocene ice volumes and sea level lowering. *J. Geol.*, Vol. 70, 206–214.

DONOVAN, D. T., HORTON, A. and IVIMEY-COOK, H. C. 1979. The transgression of the Lower Lias over the northern flank of the London Platform. *J. Geol. Soc. London*, Vol. 136, 165–173.

DUGDALE, W. 1662. *The history of imbanking and drayning of divers fennes and marches, both in foreign parts, and in this kingdom; and of the improvements thereby.* (London.)

EDMONDS, E, A. and DINHAM, C. H. 1965. Geology of the country around Huntingdon and Biggleswade. *Mem. Geol. Surv. G.B.*.

FISHER, O. 1868. On Roslyn or Roswell Hill Clay-pit, near Ely. *Geol. Mag.*, Vol. 5, 407–411.

— 1871. On supposed thermal springs in Cambridgeshire. *Geol. Mag.*, Vol. 8, 42.

— 1873. On the phosphatic nodules of the Cretaceous rocks of Cambridgeshire. *Q. J. Geol. Soc. London*, Vol. 29, 52–63.

FITTON, W. H. 1836. Observations on some of the strata between the Chalk and the Oxford Oolite in the south-east of England. *Trans. Geol. Soc. London*, Vol. 4, 103–389.

FORBES, C. L. 1960. Field meeting in the Cambridge district. *Proc. Geol. Assoc.*, Vol. 71, 233–241.

— 1965. Geology and ground-water. 1–17 in *The Cambridge region*. STEERS, J. A. (editor). (London: British Association for the Advancement of Science.)

FORBES-LESLIE, W. 1917a. The Norfolk oil-shales. *J. Inst. Petrol. Technol*, Vol. 3, 3–35.

— 1917b. The occurence of petroleum in England. *J. Inst. Petrol. Technol.*, Vol. 3, 152–190.

FOWLER, G. 1932. Old river-beds in the fenlands. *Geogr. J.*, Vol. 79, 210–12.

— 1933. Fenland waterways, past and present. South Level District. Part 1. *Proc. Cambridge. Antiq. Soc.*, Vol. 33, 108–28.

— 1934a. Fenland waterways, past and present. South Level District. Part II. *Proc. Cambridge. Antiq. Soc.*, Vol. 34, 17–33.

— 1934b. The extinct waterways of the fens. *Geogr. J.*, Vol. 83, 30–39.

— 1947. An extinct East Anglian lake. *East Anglian Mag.*, Vol. 17, 31–33.

— LETHBRIDGE, T. C. and SAYCE, R. U. 1931. A skeleton of the Early Bronze Age found in the fens. *Proc. Prehist. Soc. East Anglia*, Vol. 6, 362–364.

FOX, C. 1923. *The archaeology of the Cambridge Region*. (Cambridge: Cambridge University Press.)

GALLOIS, R. W. 1976. Ely project. 10–19, in IGS Boreholes 1975. *Rep. Inst. Geol. Sci.*, No. 76/10.

— 1978. A pilot study of oil shale occurrences in the Kimmeridge Clay. *Rep. Inst. Geol. Sci.*, No. 78/13.

— 1979a. Geological investigations for the Wash Water Storage Scheme. *Rep. Inst. Geol. Sci.*, No. 78/19.

— 1979b. The Pleistocene history of west Norfolk. *Bull. Geol. Soc. Norfolk*, Vol. 30, 3–38.

— 1984. The late Jurassic to mid Cretaceous rocks of Norfolk. *Bull. Geol. Soc. Norfolk*, Vol. 34, 3–64.

— and COX, B. M. 1974. Stratigraphy of the Upper Kimmeridge Clay of the Wash area. *Bull. Geol. Surv. G.B.*, No. 47, 1–28.

— — 1976. The stratigraphy of the Lower Kimmeridge Clay of eastern England. *Proc. Yorkshire Geol. Soc.*, Vol. 41, 13–26.

— — 1977. The stratigraphy of the Middle and Upper Oxfordian sediments of Fenland. *Proc. Geol. Assoc.*, Vol. 88, 207–228.

— and MORTER, A. A. 1976. The Trunch Borehole. 8–10 in IGS Boreholes 1975, *Rep. Inst. Geol. Sci.*, No. 76/10.

— — 1982. The stratigraphy of the Gault of East Anglia. *Proc. Geol. Assoc.*, Vol. 93, 351–368.

— and WORSSAM, B. C. 1983. Stratigraphy of the Harwell boreholes. *Rep. Fluid Processes Res. Group, Br. Geol. Surv.*, FLPU 83–14.

GODWIN, H. 1938a. The botany of Cambridgeshire. 44–59 in *A scientific survey of the Cambridgeshire District*. DARBY, H. C. (editor). (London: British Association of the Advancement of Science.)

— 1938b. The origin of roddons. *Geogr. J.*, Vol. 91, 241–250.

— 1940. Studies in the post-glacial history of British vegetation. Parts 3 and 4. *Phil. Trans. R. Soc.* Ser. B., Vol. 230, 239–304.

— 1960. Radiocarbon dating and Quaternary history in Britain. *Proc. R. Soc.* Ser. B., Vol. 153, 287–320.

— 1978. *Fenland: its ancient past and uncertain future*. (Cambridge: Cambridge University Press.)

— and CLIFFORD, M. H. 1938. Studies of the post-glacial history of British vegetation. Parts 1 and 2. *Phil. Trans. R. Soc.*, Ser. B., Vol. 229, 323–406.

— GODWIN, M. E. and CLIFFORD, M. H. 1935. Controlling factors in the formation of fen deposits, as shown by peat investigation at Wood Fen, near Ely. *J. Ecol.*, Vol. 23, 509–535.

GROVE, R. 1976. *The Cambridgeshire coprolite mining rush.* (Cambridge: Oleander Press.)

GUNN, J. 1866. Geology of Norfolk. *Geol. Mag.*, Vol. 3, 258.

HAILSTONE, J. 1816. Outlines of the geology of Cambridgeshire. *Trans. Geol. Soc.*, Vol. 3, 243.

HANCOCK, J. M. 1954. A new Ampthill Clay fauna from Knapwell, Cambridgeshire. *Geol. Mag.*, Vol. 91, 249–251.

— 1972. Crétacé: Écosse, Angleterre, Pays de Galles. Vol. 1. *Fasc. 3a XI, LEX. Stratig. Internat.* (Paris: Centre National de la Récherche Scientifique.)

— 1976. The petrology of the Chalk. *Proc. Geol. Assoc.*, Vol. 86, 499–535.

— and KAUFFMAN, E. G. 1979. The great transgressions of the Late Cretaceous. *J. Geol. Soc. London*, Vol. 136, 175–186.

HARMER, F. W. 1871. On some thermal springs in the Fens of Cambridgeshire. *Geol. Mag.*, Vol. 8, 143.

— 1877. *The testimony of the rocks in Norfolk.* (London.)

HARRISON, R. K., JEANS, C. V. and MERRIMAN, R. 1979. Mesozoic igneous rocks, hydrothermal mineralisation and volcanigenic sediments in Britain and adjacent regions. *Bull. Geol. Surv. G.B.*, No. 70, 57–69.

HART, M. B. 1973. Foraminiferal evidence for the age of the Cambridge Greensand. *Proc. Geol. Assoc.*, Vol. 84, 65–82.

HAWKES, L. 1943. The erratics of the Cambridge Greensand — their nature, provenance and mode of transport. *Q. J. Geol. Soc. London*, Vol. 99, 93–104.

HODGE, C. A. H. and SEALE, R. S. 1966. The soils of the district around Cambridge. *Mem. Soil Surv. G.B.*.

HOLLAND, C. H. and OTHERS. 1978. A guide to stratigraphical procedure. *Spec. Rep. Geol. Soc. London*, No. 11.

HOLMES, S. C. A., TAYLOR, J. H., EARP, J. R. and WORSSAM, B. C. 1965. One-Inch Geological Sheet 188 (Cambridge). *Geol. Surv. G.B.*.

HORTON, A., LAKE, R. D., BISSON, G. and COPPACK, B. C. 1974. The geology of Peterborough. *Rep. Inst. Geol. Sci.*, No. 73/12.

— and POOLE, E. G. 1977. The lithostratigraphy of three geophysical marker horizons in the Lower Lias of Oxfordshire. *Bull. Geol. Surv. G.B.*, No. 62, 13–24.

HOUBOLT, J. H. C. 1968. Recent sediments in the southern bight of the North Sea. *Geol. Mijnbouwkd*, Vol. 47, 245–273.

HOUSE, M. R., RICHARDSON, J. B., CHALONER, W. G., ALLEN, J. R. L., HOLLAND, C. H. and WESTOLL, T. S. 1977. A correlation of the Devonian rocks of the British Isles. *Spec. Rep. Geol. Soc. London*, No. 7.

HUDSON, J. D. 1963. The recognition of salinity-controlled mollusc assemblages in the Great Estuarine Series (Middle Jurassic) of the Inner Hebrides. *Palaeontology*, Vol. 6, 318–326.

— 1980. Aspects of brackish-water facies and faunas from the Jurassic of north-west Scotland. *Proc. Geol. Assoc.*, Vol. 91, 99–105.

HUTCHINSON, J. N. 1980. The record of peat wastage in the East Anglian fenlands at Holme Post, 1848–1978 A.D. *J. Ecol.*, Vol. 68, 229–249.

IMBRIE, J., HAYS, J. D., MARTINSON, D. G., McINTYRE, A., MIX, A. C., MORLEY, J. J., PISIAS, N. G., PRELL, W. L. and SHACKLETON, N. J. 1984. The orbital theory of Pleistocene climate: support from a revised chronology of the marine $\delta^{18}O$

record. 269–305 in *Milankovitch and Climate*, Part 1. BERGER, A. L., IMBRIE, J., HAYS, J., KUKLA, G. and SALTZMAN, B (editors). (Dordrecht: D. Reidol.)

INSTITUTE OF GEOLOGICAL SCIENCES. 1966. *Annual report for 1965.* (London: Institute of Geological Sciences.)

— 1971. *Annual report for 1970.* (London: Institute of Geological Sciences.)

— 1972. *Annual report for 1971.* (London: Institute of Geological Sciences.)

JEANS, C. V. 1968. The origin of the montmorillonite of the European Chalk with special reference to the Lower Chalk of England. *Clay Minerals*, Vol. 7, 311–330.

— 1978. Silicifications and associated clay assemblages in the Cretaceous marine sediments of southern England. *Clay Minerals*, Vol. 13, 101–126.

JENNINGS, J. N. 1950. The origin of the fenland meres: Fenland homologues of the Norfolk Broads. *Geol. Mag.*, Vol. 87, 217–225.

JONAS, S. 1847. On the farming of Cambridgeshire. *J. R. Agric. Soc.*, Vol. 17, 35–72.

JUKES-BROWNE, A. J. 1875. On the relations of the Cambridge Gault and Greensand. *Q. J. Geol. Soc. London*, Vol. 31, 256–316.

— 1900. Cretaceous rocks of Britain, Vol. 1, The Gault and Upper Greensand of England. *Mem. Geol. Surv. G.B.*

— 1903. The Cretaceous rocks of Britain. Vol. 2, The Lower and Middle Chalk. *Mem. Geol. Surv. G.B.*

— and HILL, W. 1887. On the lower part of the Upper Cretaceous Series in West Suffolk and Norfolk. *Q. J. Geol. Soc. London*, Vol. 43, 544–597.

KEEPING, W. 1883. *The fossils and palaeontological affinities of the Neocomian deposits of Upware and Brickhill.* (Cambridge: Cambridge University Press.)

KENNEDY, W. J. 1970. Trace fossils in the Chalk environment. 263–282 in *Trace fossils.* CRIMES, T. P. and HARPER, J. C. (editors). *Geol. J. Spec. Issue.* No. 3.

— and HANCOCK, J. M. 1978. The mid-Cretaceous of the United Kingdom. V 1 to V 72 *in* Mid-Cretaceous events: reports of the biostratigraphy of key areas. *Ann. Mus. Hist. Nat. Nice.*, Vol. IV.

KENT, P. E. 1947. A deep boring at North Creake, Norfolk. *Geol. Mag.*, Vol. 84, 2–18.

— 1962. A borehole to basement rocks at Glinton, near Peterborough, Northants. *Proc. Geol. Soc. London.* No. 1595. 40–42.

— 1968. The buried floor of Eastern England. 138–148 in *The geology of the East Midlands.* SYLVESTER-BRADLEY, P. C. and FORD, T. D. (editors). (Leicester: Leicester University Press.)

KERNEY, M. P. 1963. Late-glacial deposits in the Chalk of south-east England. *Phil. Trans. R. Soc.*, Ser. B, Vol. 246, 203–254.

LARWOOD, G. P. and FUNNELL, B. M. (editors) 1961. The geology of Norfolk. *Trans. Norfolk Norwich Nat. Soc.*, Vol. 19, 270–375.

LINSSER, H. 1968. Transformation of magnetometric data into tectonic maps by Digital Template Analysis. *Geophys. Prospect.*, Vol. 16, 179–207.

LUNN, F. 1819. On the strata of the northern division of Cambridgeshire. *Trans. Geol. Soc.*, Vol. 5, 114.

MCKENNY HUGHES, T. 1884. Excursion to Cambridge. *Proc. Geol. Assoc.*, Vol. 8, 399–404.

— 1894. Excursion to Cambridge and Ely. *Proc. Geol. Assoc.*, Vol. 13, 292–295.

MARR, J. E. 1926. The Pleistocene deposits of the lower part of the Great Ouse Basin. *Q. J. Geol. Soc. London*, Vol. 82, 101–143.

MEDD, A. W. 1979. The Upper Jurassic coccoliths from the Haddenham and Gamlingay boreholes (Cambridgeshire, England). *Eclogae Geol. Helv.*, Vol. 72, 19–109.

MIDDLEMISS, F. A. 1967. Analysis of structure in a region of gentle en echelon folding. *Neues Jahr. Miner. Geol. Paläont. Abh.*, Vol. 129, 137–156.

— 1976. Studies of the sedimentation of the Lower Greensand of the Weald, 1875–1975: a review and commentary. *Proc. Geol. Assoc.*, Vol. 86 for 1975, 457–473.

MILLER, S. H. and SKERTCHLY, S. B. J. 1878. *The Fenland past and present.* (London: Longmans Green.)

MITCHELL, G. F., PENNY, L. F., SHOTTON, F. W. and WEST, R. G. 1973. A correlation of Quaternary deposits in the British Isles. *Spec. Rep. Geol. Soc. London*, No. 4.

MOORE, J. 1685. *Map of the Bedford Level.* (MS: Cambridge University Library.)

MORTIMER, M. G. 1967. Some Lower Devonian microfloras from southern Britain. *Rev. Palaeobot. Palynol.*, Vol. 1, 95–109.

— and CHALONER, W. G. 1972. The palynology of the concealed Devonian rocks of southern England. *Bull. Geol. Surv. G.B.*, No. 39, 1–56.

NARAYAN, J. 1963. Cross-stratification and palaeogeography of the Lower Greensand of south-east England and Bas-Boulonnais, France. *Nature, London*, Vol. 199, 1246–7.

— 1971. Sedimentary structures in the Lower Greensand of the Weald, England, and Bas-Boulonnais, France. *Sediment. Geol.*, Vol. 6, 73.–109.

OAKLEY, K. P. 1941. British phosphates. *Wartime Pamph., Geol. Surv. G.B.*, No. 8, Pt. 3.

OSBORNE WHITE, H. J. O. 1932. The geology of the country around Saffron Walden. *Mem. Geol. Surv. G.B.*.

OWEN, H. G. 1971. Middle Albian stratigraphy in the Anglo-Paris Basin. *Bull. Br. Mus. (Nat. Hist). Geol.*, Supplement 8.

— 1975. The stratigraphy of the Gault and Upper Greensand of the Weald. *Proc. Geol. Assoc.*, Vol. 86, 475–498.

— 1979. Ammonite zonal stratigraphy in the Albian of North Germany and its setting in the Hoplitinid Faunal Province. *Aspekte der Kreide Europas*, IUGS Series A, Vol. 6, 563–588.

OWEN, E. F., RAWSON, P. F. and WHITHAM, F. 1968. The Carstone (Lower Cretaceous) of Melton, East Yorkshire, and its brachiopod fauna. *Proc. Yorkshire Geol. Soc.*, Vol. 36, 513–524.

PARRY, R. H. G. 1972. Some properties of heavily overconsolidated Oxford Clay at a site near Bedford. *Géotechnique*, Vol. 22, 485–507.

PEAKE, N. B. and HANCOCK, J. M. 1961. The Upper Cretaceous of Norfolk. *Trans. Norfolk Norwich Nat. Soc.*, Vol. 19, 293–339.

PENNING, W. H. and JUKES-BROWNE, A. J. 1881. Geology of the neighbourhood of Cambridge. *Mem. Geol. Surv. G.B.*.

PERRIN, R. M. S. 1957. The clay mineralogy of some tills in the Cambridge district. *Clay Minerals Bull.*, Vol. 3, 193–205.

— 1971. *The clay mineralogy of British sediments.* (London: Mineralogical Society.)

PHILLIPS, C. W. (editor). 1970. *The Fenland in Roman times.* (London: The Royal Geographical Society.)

PREECE, R. C. and VENTRIS, P. A. 1983. An interglacial site at Galley Hill, near St. Ives, Cambridgeshire. *Bull. Geol. Soc. Norfolk*, No. 33, 63–72.

PRICE, F. G. H. 1874. On the Gault of Folkestone. *Geol. Soc. London*, Vol. 30, 342–368.

— 1875. On the Lower Greensand and Gault of Folkestone. *Proc. Geol. Assoc.*, Vol. 4, 135–50.

— 1879. *The Gault.* (London: Taylor and Francis.)

PRINGLE, J. 1923. On the concealed Mesozoic rocks in south-west Norfolk. *Summ. Prog. Geol. Surv. G.B. for 1922*, 126–139.

RASTALL, R. H. 1910. Cambridgeshire, Bedfordshire and West Norfolk. 124–178 in *Geology in the field: Jubilee Vol. Geol. Assoc.*

— 1919. The mineral composition of the Lower Greensand strata of eastern England. *Geol. Mag.*, Vol. 56, 211–20 and 265–72.

— 1925. On the tectonics of the southern Midlands. *Geol. Mag.*, Vol. 62, 193–222.

RAWSON, P. F., CURRY, D., DILLEY, F. C., HANCOCK, J. M., KENNEDY, W. J., NEALE, J. W., WOOD, C. J., and WORSSAM, B. C. 1978. A correlation of the Cretaceous rocks in the British Isles. *Spec. Rep. Geol. Soc. London*, No. 9.

REED, F. R. C. 1897. *A handbook to the geology of Cambridgeshire.* (Cambridge: Cambridge University Press.)

ROBERTS, T. 1892. *The Jurassic rocks of the neighbourhood of Cambridge. Sedgwick Prize Essay for 1886.* (Cambridge: Cambridge University Press.)

ROSE, C. B. 1835–36. A sketch of the geology of west Norfolk. *Phil. Mag.*, Vol. 7, 171–182, 274–279, 370–376; Vol. 8, 28–42.

— 1859. Geological pearls. *Geologist*, Vol. 2, 295

— 1862. On the Cretaceous group in Norfolk. *Proc. Geol. Assoc.*, Vol. 1, 234–236.

SALWAY, P. 1970. The Fenland. 1–21 in *The Fenland in Roman times.* C. W. PHILLIPS (editor). (London: The Royal Geographical Society.)

SAMUELS, S. G. 1975. Some properties of the Gault Clay from the Ely–Ouse Essex Water tunnel. *Géotechnique*, Vol. 25, 239–264.

SCHWARZACHER, W. 1953. Cross-bedding and grain size in the Lower Cretaceous sands of East Anglia. *Geol. Mag.*, Vol. 90, 322–330.

SEALE, R. S. 1956. The heavy minerals of some soils from the neighbourhood of Cambridge, England. *J. Soil Sci.*, Vol. 7, 307–318.

— 1974. Geology of the Ely district. *Bull. Geol. Soc. Norfolk*, Vol. 25, 21–36.

— 1975. Soils of the Ely district. *Mem. Soil Surv. G.B.*

SEDGWICK, A. 1846. On the geology of the neighbourhood of Cambridge, including the formations between the Chalk escarpment and the Great Bedford Level. *Rep. Br. Assoc. [for 1845]*, 40.

SEELEY, H. G. 1861. Notice of the Elsworth and other new rocks in the Oxford Clay, and of the Bluntisham Clay above them. *Geologist*, Vol. 4, 460–461.

— 1865a. On the significance of the sequence of rocks and fossils: theoretical considerations on the Upper Secondary rocks, as seen in the section at Ely. *Geol. Mag.*, Vol. 2, 262–265.

— 1865b. On a section discovering the Cretaceous beds at Ely. *Geol. Mag.,* Vol. 2, 529–534.

— 1866a. A sketch of the gravels and drift of the Fenland. *Q. J. Geol. Soc. London,* Vol. 22, 470–480.

— 1866b. The rock of the Cambridge Greensand. *Geol. Mag.,* Vol. 3, 302–307.

— 1868. On the collocation of the strata at Roswell Hole, near Ely. *Geol. Mag.,* Vol. 5, 347–349.

— 1869. *Index to the fossil remains of Aves, Ornithosauria and Reptilea from the Secondary System of strata, arranged in the Woodwardian Museum of The University of Cambridge.* (Cambridge: Cambridge University Press.)

SHACKLETON, N. J. and OPDYKE, N. D. 1973. Oxygen isotope and palaeomagnetic stratigraphy of Equatorial Pacific Core V28–238: oxygen isotope temperatures and ice volumes on a 10^5 year and 10^6 year scale. *Quaternary Res.,* Vol. 3, 39–55.

SKERTCHLY, S. B. J. 1877. The geology of Fenland. *Mem. Geol. Surv. G.B..*

SMITH, W. 1815. *A delineation of the strata of England and Wales, with part of Scotland.* (W. Smith.)

— 1819a. *Geological map of Cambridgeshire.* (London: J. Carey.)

— 1819b. *Geological map of Norfolk.* (London: J. Carey.)

SPARKS, B. W. and WEST, R. G. 1965. The relief and drift deposits. 18–40 in *The Cambridge region.* J. A. STEERS (editor). (London: British Association for the Advancement of Science.)

—·— 1970. Late Pleistocene deposits at Wretton, Norfolk. I. Ipswichian interglacial deposits. *Proc. R. Soc.,* Ser. A, Vol. 258, 1–30.

— WILLIAMS, R. B. G. and BELL, F. G. 1972. Presumed ground-ice depressions in East Anglia. *Proc. R. Soc.,* Ser. A, Vol. 327, 329–343.

SPATH, L. F. 1923–43. A monograph of the ammonoidea of the Gault. *Palaeontogr. Soc. Monogr.,* 2 Vols., 1–787.

— 1924. On the ammonites of the Specton Clay and the subdivisions of the Neocomian. *Geol. Mag.,* Vol. 61, 73–89.

— 1939. The ammonite zones of the Upper Oxford Clay of Warboys, Huntingdonshire. *Bull. Geol. Surv. G.B.* No. 1, 82–98.

STRAW, A. 1960. The limit of the 'last' glaciation in north Norfolk. *Proc. Geol. Assoc.* Vol. 71, 379–390.

— 1979. The geomorphological significance of the Wolstonian glaciation of eastern England. *Trans. Inst. Br. Geogr.,* Vol. 4, 540–549.

STRIDE, A. H. 1963. Current-swept sea floors near the southern half of Great Britain. *Q. J. Geol. Soc. London,* Vol. 119, 175–99.

SUGGATE, R. P. and WEST, R. G. 1959. On the extent of the last glaciation in eastern England. *Proc. R. Soc.,* Ser. B, Vol. 150, 263–283.

SYKES, R. M. and CALLOMON, J. H. 1979. The *Amoeboceras* zonation of the Boreal Upper Oxfordian. *Palaeontology,* Vol. 22, 839–903.

SYLVESTER-BRADLEY, P. C. and FORD, T. D. (editors). 1968. *The geology of the East Midlands.* (Leicester: Leicester University Press.)

SZABO, B. J. and COLLINS, D. 1975. Ages of fossil bone from British interglacial sites. *Nature, London,* Vol. 254, 680–682.

TARLO, L. B. 1958. The scapula of *Pliosaurus macromerus* Phillips. *Palaeontology,* Vol. 1, 193–199.

— 1959a. *Stretosaurus* gen. nov., a giant pliosaur from the Kimmeridge Clay. *Palaeontology,* Vol. 2, 39–55.

— 1959b. *Pliosaurus brachyspondylus* (Owen) from the Kimmeridge Clay. *Palaeontology,* Vol. 1, 283–291.

TEALL, J. J. H. 1875. *The Potton and Wicken phosphatic deposits. Sedgwick Prize Essay for 1873.* (Cambridge: Cambridge University Press.)

TOPLEY, W. 1875. The geology of the Weald. *Mem. Geol. Surv. G.B.*

TRIMMER, J. 1846. On the geology of Norfolk. *J. R. Agric. Soc.,* Vol. 7, 449.

VERMUYDEN, C. 1642. *A discourse touching the drayning the great fennes etc.* (London.)

WARD, W. H., BURLAND, J. B. and GALLOIS, R. W. 1968. Geotechnical assessment of a site at Mundford, Norfolk, for a large proton accelerator. *Géotechnique,* Vol. 18, 399–431.

WATTS, A. S., PERRIN, R. M. S. and WEST, R. G. 1966. Patterned ground in the Breckland; structure and composition. *J. Ecol.,* Vol. 54, 239–258.

WEDD, C. B. 1898. On the Corallian rocks of Upware (Cambs.). *Q. J. Geol. Soc. London,* Vol. 54, 601–619.

WELLS, C. 1964. Pathological epipodials and tarsus in *Stretosaurus macromerus* from the Kimmeridge Clay, Stretham, Cambridgeshire. *Q. J. Geol. Soc. London,* Vol. 120, 299–304.

WELLS, S. 1830. *The history of the drainage of the great level of the fens called Bedford Level.* (London.)

WENTWORTH-DAY, J. 1954. *A history of the fens.* (London: Harrap.)

WEST, R. G. and DONNER, J. J. 1956. The glaciations of East Anglia and the East Midlands. *Q. J. Geol. Soc. London,* Vol. 111, 69–91.

— DICKSON, C. A., CATT, J. A., WEIR, A. H. and SPARKS, B. W. 1974. Late Pleistocene deposits at Wretton, Norfolk. 2 Devensian deposits. *Proc. R. Soc. Ser. B,* Vol. 267, 337–420.

WHITAKER, W. 1883. Excursion to Hunstanton. *Proc. Geol. Assoc.,* Vol. 8, 133.

— 1906. The water supply of Suffolk from underground sources. *Mem. Geol. Surv. G.B..*

— 1921. The water supply of Norfolk from underground sources. *Mem. Geol. Surv. G.B.*

— 1922. The water supply of Cambridgeshire, Huntingdon and Rutland, from underground sources. *Mem. Geol. Surv. G.B.*

— WOODWARD, H. B., BENNETT, F. J., SKERTCHLY, S. B. J. and JUKES-BROWNE, A. J. 1891. The geology of parts of Cambridgeshire and Suffolk (Ely, Mildenhall and Thetford). *Mem. Geol. Surv. G.B..*

— SKETCHLEY, S. B. J. and JUKES-BROWNE, A. J. 1893. The geology of the south-western and northern parts of Cambridgeshire. *Mem. Geol. Surv. G.B..*

— and JUKES-BROWNE, A. J. 1899. The geology of the borders of The Wash. *Mem. Geol. Surv. G.B..*

WILLIAMS, R. B. G. 1964. Fossil patterned ground in eastern England. *Biul. Peryglac.,* Vol. 14, 337–349.

WILLIS, E. H. 1961. Marine transgression sequences in the English Fenlands. *Ann. New York Acad. Sci.,* Vol. 95, 368–376.

WILLS, L. J. 1951. *A palaeogeographical atlas.* (Glasgow: Blackie and Son).

— 1973. A palaeogeologic map of the Palaeozoic floor below the Upper Permian and Mesozoic formations. *Mem. Geol. Soc. London,* No. 7.

WOODLAND, A. W. 1943. Water supply from underground sources of Cambridge-Ipswich district. Part 2. Well-catalogues for new series One-Inch Sheets 173 (Ely), 188 (Cambridge) and 205 (Saffron Walden). *Wartime Pamph. Geol. Surv. G.B.*, No. 20.

— 1970. The buried tunnel-valleys of East Anglia. *Proc. Yorkshire Geol. Soc.*, Vol. 37, 521–578.

WOODWARD, A. S. 1890. On the head of *Eurycormus* from the Kimmeridge Clay of Ely. *Geol. Mag.*, Vol. 7, 289–292.

WOODWARD, H. B. 1895. The Jurassic rocks of Britain, Vol. 5, The Middle and Upper Oolitic rocks. *Mem. Geol. Surv. G.B.*

WOODWARD, S. 1833. *An outline of the geology of Norfolk.* (Norwich.)

WORSSAM, B. C. and TAYLOR, J. H. 1969. Geology of the country around Cambridge. *Mem. Geol. Surv. G.B.*

WYATT, R. J. and others. 1984. 1:50 000 Geological Sheet 158 (Peterborough). *Geol. Surv. G.B.*

ZIEGLER, B. 1962. Die Ammoniten-Gattung *Aulacostephanus* im Oberjura (Taxonomie, Stratigraphie, Biologie). *Palaeontographica*, Vol. 119A, 1–172.

FOSSIL INDEX

GENERAL INDEX

BRITISH GEOLOGICAL SURVEY

Keyworth, Nottingham NG12 5GG

Murchison House, West Mains Road,
Edinburgh EH9 3LA

The full range of Survey publications is available
through the Sales Desks at Keyworth and
Murchison House. Selected items are stocked by
the Geological Museum Bookshop, Exhibition
Road, London SW7 2DE; all other items may be
obtained through the BGS London Information
Office in the Geological Museum. All the books
are listed in HMSO's Sectional List 45. Maps are
listed in the BGS Map Catalogue and Ordnance
Survey's Trade Catalogue. They can be bought
from Ordnance Survey Agents as well as from
BGS.

*The British Geological Survey carries out the geological
survey of Great Britain and Northern Ireland (the latter as
an agency service for the government of Northern Ireland),
and of the surrounding continental shelf, as well as its
basic research projects. It also undertakes programmes of
British technical aid in geology in developing countries as
arranged by the Overseas Development Administration.*

*The British Geological Survey is a component body of the
Natural Environment Research Council.*

Maps and diagrams in this book use topography
based on Ordnance Survey mapping

HER MAJESTY'S STATIONERY OFFICE

HMSO publications are available from:

HMSO Publications Centre
(Mail and telephone orders)
PO Box 276, London SW8 5DT
Telephone orders (01) 622 3316
General enquiries (01) 211 5656
Queueing system in operation for both numbers

HMSO Bookshops
49 High Holborn, London WC1V 6HB
  (01) 211 5656 (Counter service only)
258 Broad Street, Birmingham B1 2HE
  (021) 643 3740
Southey House, 33 Wine Street, Bristol BS1 2BQ
  (0272) 264306
9 Princess Street, Manchester M60 8AS
  (061) 834 7201
80 Chichester Street, Belfast BT1 4JY
  (0232) 238451
71 Lothian Road, Edinburgh EH3 9AZ
  (031) 228 4181

HMSO's Accredited Agents
(see Yellow Pages)

And through good booksellers